SpringerBriefs in Electrical and Computer Engineering

Computational Electromagnetics

Series editor

Rakesh Mohan Jha, Bangalore, India

More information about this series at http://www.springer.com/series/13885

Hema Singh · Simy Antony
Rakesh Mohan Jha

Plasma-based Radar
Cross Section Reduction

 Springer

Hema Singh
Centre for Electromagnetics
CSIR-National Aerospace Laboratories
Bangalore, Karnataka
India

Rakesh Mohan Jha
Centre for Electromagnetics
CSIR-National Aerospace Laboratories
Bangalore, Karnataka
India

Simy Antony
Centre for Electromagnetics
CSIR-National Aerospace Laboratories
Bangalore, Karnataka
India

ISSN 2191-8112   ISSN 2191-8120   (electronic)
SpringerBriefs in Electrical and Computer Engineering
ISSN 2365-6239   ISSN 2365-6247   (electronic)
SpringerBriefs in Computational Electromagnetics
ISBN 978-981-287-759-8   ISBN 978-981-287-760-4   (eBook)
DOI 10.1007/978-981-287-760-4

Library of Congress Control Number: 2015947122

Springer Singapore Heidelberg New York Dordrecht London

Printed on acid-free paper

Springer Science+Business Media Singapore Pte Ltd. is part of Springer Science+Business Media
(www.springer.com)

To Professor R. Narasimha

In Memory of Dr. Rakesh Mohan Jha
Great scientist, mentor, and excellent
human being

Dr. Rakesh Mohan Jha was a brilliant contributor to science, a wonderful human being, and a great mentor and friend to all of us associated with this book. With a heavy heart we mourn his sudden and untimely demise and dedicate this book to his memory.

Foreword

National Aerospace Laboratories (NAL), a constituent of the Council of Scientific and Industrial Research (CSIR), is the only civilian aerospace R&D Institution in India. CSIR-NAL is a high-technology institution focusing on various disciplines in aerospace and has a mandate to develop aerospace technologies with strong science content, design and build small and medium-size civil aircraft prototypes, and support all national aerospace programs. It has many advanced test facilities including trisonic wind tunnels which are recognized as national facilities. The areas of expertize and competencies include computational fluid dynamics, experimental aerodynamics, electromagnetics, flight mechanics and control, turbomachinery and combustion, composites for airframes, avionics, aerospace materials, structural design, analysis, and testing. CSIR-NAL is located in Bangalore, India, with the CSIR Headquarters being located in New Delhi.

CSIR-NAL and Springer have recently signed a cooperation agreement for the publication of selected works of authors from CSIR-NAL as Springer book volumes. Within these books, recent research in the different fields of aerospace that demonstrate CSIR-NAL's outstanding research competencies and capabilities to the global scientific community will be documented.

The first set of five books are from selected works carried out at the CSIR-NAL's Centre for Electromagnetics, and are presented as part of the series SpringerBriefs in Computational Electromagnetics, which is a sub-series of SpringerBriefs in Electrical and Computer Engineering.

CSIR-NAL's Centre for Electromagnetics mainly addresses issues related to electromagnetic (EM) design and analysis carried out in the context of aerospace engineering in the presence of large airframe structures, which is vastly different and in contrast to classical electromagnetics and which often assumes a free-space ambience. The pioneering work done by the Centre for Electromagnetics in some of these niche areas has led to founding the basis of contemporary theories. For example, the geodesic constant method (GCM) proposed by the scientists of the Centre for Electromagnetics is immensely popular with the peers worldwide, and forms the basis for modern conformal antenna array theory.

The activities of the Centre for Electromagnetics consist of (i) Surface modeling and ray tracing, (ii) Airborne antenna analysis and siting (for aircraft, satellites and SLV), (iii) Radar cross-section (RCS) studies of aerospace vehicles, including radar absorbing materials (RAM) and structures (RAS), RCS reduction and active RCS reduction, (iv) Phased antenna arrays, conformal arrays, and conformal adaptive array design, (v) Frequency-selective surface (FSS), (vi) Airborne and ground-based radomes, (vii) Metamaterials for aerospace applications including in the Terahertz (THz) domain, and (viii) EM characterization of materials.

It is hoped that this dissemination of information through these SpringerBriefs will encourage new research as well as forge new partnerships with academic and research organizations worldwide.

Shyam Chetty
Director
CSIR-National Aerospace Laboratories
Bangalore, India

Preface

The reduction and control of radar cross section (RCS) of an object have been attempted by various techniques, including shaping, use of radar absorbing materials, frequency-selective surfaces, engineered materials etc. Likewise, plasma-based stealth is a RCS reduction technique associated with the reflection and absorption of incident EM wave by the plasma layer surrounding the structure. This book reviews the research and development work on plasma-based RCS reduction reported so far in the open literature. The book starts with the basics of EM wave interactions with plasma, briefly discusses the methods used to analyze the propagation characteristics of plasma, and the plasma generation. It presents the parametric analysis of propagation behavior of plasma, and the challenges in the implementation of plasma-based stealth technology. The book provides an insight of role of various parameters in the EM propagation within plasma. This comprehensive review is expected to serve as a parametric base for researchers working in the area of plasma stealth.

Hema Singh
Simy Antony
Rakesh Mohan Jha

Acknowledgments

We would like to thank Mr. Shyam Chetty, Director, CSIR-National Aerospace Laboratories, Bangalore for his permissions and support to write this SpringerBrief.

We would also like to acknowledge valuable suggestions from our colleagues at the Centre for Electromagnetics, Dr. R.U. Nair, Dr. Shiv Narayan, Dr. Balamati Choudhury, and Mr. K.S. Venu and their invaluable support during the course of writing this book. We would like to thank Mr. Harish S. Rawat, Ms. P.S. Neethu, Mr. Umesh V. Sharma, and Mr. Bala Ankaiah, the project staff at the Centre for Electromagnetics, for their consistent support during the preparation of this manuscript.

But for the concerted support and encouragement from Springer, especially the efforts of Suvira Srivastav, Associate Director, and Swati Mehershi, Senior Editor, Applied Sciences and Engineering, it would not have been possible to bring out this book within such a short span of time. We very much appreciate the continued support by Ms. Kamiya Khatter and Ms. Aparajita Singh of Springer toward bringing out this brief.

Hema Singh
Simy Antony
Rakesh Mohan Jha

Contents

About the Authors

Dr. Hema Singh is currently working as a Senior Scientist in Centre for Electromagnetics of CSIR-National Aerospace Laboratories, Bangalore, India. Earlier, she was a Lecturer in EEE, BITS, Pilani, India during 2001–2004. She obtained her Ph.D. degree in Electronics Engineering from IIT-BHU, Varanasi India in 2000. Her active area of research is Computational Electromagnetics for Aerospace Applications. More specifically, the topics she has contributed to are GTD/UTD, EM analysis of propagation in an indoor environment, phased arrays, conformal antennas, radar cross-section (RCS) studies including Active RCS Reduction. She received Best Woman Scientist Award in CSIR-NAL, Bangalore for the period of 2007–2008 for her contribution in the areas of phased antenna array, adaptive arrays, and active RCS reduction. Dr. Singh has co-authored one book, one book chapter, and over 120 scientific research papers and technical reports.

Simy Antony obtained B.Tech. (ECE) from University of Calicut, India and M.Tech. in Electronics (Microwave and Radar Electronics) from Cochin University of Science and Technology, Kerala, India. She was a Project Scientist at the Centre for Electromagnetics of CSIR-National Aerospace Laboratories, Bangalore where she worked on RCS studies for aerospace vehicles.

Dr. Rakesh Mohan Jha was Chief Scientist & Head, Centre for Electromagnetics, CSIR-National Aerospace Laboratories, Bangalore. Dr. Jha obtained a dual degree in BE (Hons.) EEE and M.Sc. (Hons.) Physics from BITS, Pilani (Raj.) India, in 1982. He obtained his Ph.D. (Engg.) degree from Department of Aerospace Engineering of Indian Institute of Science, Bangalore in 1989, in the area of computational electromagnetics for aerospace applications. Dr. Jha was a SERC (UK) Visiting Post-Doctoral Research Fellow at University of Oxford, Department of Engineering Science in 1991. He worked as an Alexander von Humboldt Fellow at the Institute for High-Frequency Techniques and Electronics of the University of Karlsruhe, Germany (1992–1993, 1997). He was awarded the Sir C.V. Raman Award for Aerospace Engineering for the Year 1999. Dr. Jha was elected Fellow of INAE in 2010, for his contributions to the EM Applications to Aerospace Engineering. He was also the Fellow of IETE and Distinguished Fellow of ICCES.

Dr. Jha has authored or co-authored several books, and more than five hundred scientific research papers and technical reports. He passed away during the production of this book of a cardiac arrest.

List of Figures

List of Tables

Plasma-based Radar Cross Section Reduction

Abstract The concealment of aircraft from radar sources or stealth is achieved either through shaping, or radar absorbing coatings, or engineered materials, or plasma, etc. Plasma-based stealth is a radar cross-section (RCS) reduction technique associated with the reflection and absorption of incident electromagnetic (EM) wave by the plasma layer surrounding the structure. Plasma cloud covering the aircraft may give rise to other signatures such as thermal, acoustic, infrared, or visual. Thus it is a matter of concern that the RCS reduction by plasma enhances its detectability due to other signatures. This needs a careful approach toward the plasma generation and its EM wave interaction. This book presents a comprehensive review of the plasma-based stealth, covering the basics, methods, parametric analysis, and challenges toward the realization of the idea. The book starts with the basics of EM wave interactions with plasma, briefly discusses the methods used to analyze the propagation characteristics of plasma, and its generation. It presents the parametric analysis of propagation behavior of plasma, and the challenges in the implementation of plasma-based stealth technology. This review serves as a starting point for graduate and research students, scientists, and engineers working in the area of low-observables and stealth technology.

1 Introduction

Stealth or low observability does not imply complete disappearance from the radar sources. Rather, it means low radar signature, i.e., the target is detected and tracked at a shorter distance from the radar. This is similar to camouflage tactics used by the soldiers in terrestrial warfare. Unless the soldier comes near to the enemy, they cannot detect him. Generally, the stealth technology includes everything that minimizes the signatures and prevents the detection and identification of the target (Vass 2003).

© The Author(s) 2016
H. Singh et al., *Plasma-based Radar Cross Section Reduction*,
SpringerBriefs in Computational Electromagnetics,
DOI 10.1007/978-981-287-760-4_1

In short, the importance of the RCS relies on the fact that it directly affects the detection range of the radar. This is apparent from the radar range equation (Skolnik 2003), expressed as

$$R_{\max} = 4\sqrt{\frac{P_t G A_e \sigma}{(4\pi)^2 S_{\min}}} \tag{1}$$

where $R_{\max}$ is the maximum detection range, P_t is the transmitted power, G and A_e are the gain and effective area of the transmitting and receiving antenna (which coincides for the monostatic case), σ is RCS of the target, and $S_{\min}$ is the minimum detectable signal.

The concealment of aircraft from radar sources or stealth has been practically achieved through shaping, coating, engineered materials, plasma, etc. In particular, plasma-based stealth is still a subject of research. It is conjectured that Russia and USA have implemented plasma-based low observability in the aerospace sector (http://www.rense.com/general168/newrussianstealth). However, due to strategic reasons, the details and data are not available in open domain.

In general, a stealth aircraft should be stealthy in six aspects, viz. radar, infrared, visual, acoustic, smoke, and contrail (e.g., plasma trail). Plasma stealth, also referred to as an active stealth technology, was first developed by Russia. This was made possible by placing plasma torch on the aircraft nose. This torch creates an ionized cloud around the aircraft, which absorbs the incident radar waves (Fig. 1). It has been reported in Journal of Electronic Defense, 2002 that technology for the generation of plasma cloud for stealth application has achieved RCS reduction of 20 dB in fighter bomber aircraft (Su-27IB).

The weight of the plasma generator was, however, bulky (~ 100 kg). The generated plasma shield partially consumes radar energy and causes it to bend around the aircraft, thereby reducing the radar cross section (RCS) by two orders of magnitude. Russians basically used cold plasma for stealth applications.

Plasma can be hot or cold depending on its degree of ionization. Hot plasmas are fully ionized (Dinklage 2005), whereas only a fraction of gas molecules are ionized in cold plasmas. Ionized electrons in cold plasmas do not have sufficient energy to

Fig. 1 Plasma cloud covering an aircraft

escape from its corresponding ion (i.e., do not have random nature). So the influence of temperature on the cold plasma parameters is negligible.

Plasma-based shielding is based on the fact that plasma being dispersive media absorbs the incident EM radiation before it is scattered by the target. Furthermore, the plasma–air interface, being continuous in terms of electrical dimensions, results in reduced radar signatures as compared to the target surface, which poses a sharp discontinuity for the incident wave.

Plasma cloud covering the aircraft may give rise to other signatures such as thermal, acoustic, infrared, or visual. Thus it is a matter of concern that the RCS reduction by plasma enhances its detectability due to other signatures. This needs a careful approach toward the plasma generation and its electromagnetic wave interaction. Furthermore, the parameters of plasma that can be controlled to reduce the detectability by the radar need to be identified and optimized.

This book presents a review of the plasma-based RCS reduction. Section 1.1 covers the theoretical background basics of plasma. Section 1.2 describes the plasma parameters that control the plasma performance. In Sect. 1.3, the developments in the area of plasma stealth are briefly discussed. Section 2 describes how plasma interacts with incident EM wave.

Sections 2.1 and 2.2 cover the behavior of magnetized and unmagnetized plasma, respectively. In Sect. 3, the RCS aspects of simple shapes covered with plasma are presented. Section 3.1 discusses mechanism for plasma generation and the power requirements for sustaining plasma. The approximations and methods for estimating and reducing the RCS of the target covered with plasma are discussed in Sect. 3.2. In Sect. 3.3, the performance of the plasma along with radar absorbing material is presented. The challenges in implementation of plasma-based stealth are discussed in Sect. 4. Section 5 summarizes the review on RCS reduction using plasma.

1.1 Concepts of Plasma Physics

Plasma is considered as the fourth state of matter. It has been reported that 99 % matter of the universe are in plasma state (Chen 1974). The examples of plasma include lightning bolt flash, the conducting gas within the fluorescent tube, glow of the aurora borealis, ionization in the rocket exhaust, etc.

Plasma is known to be a gas of charged particles. More specifically, it is an electrically neutral, highly ionized gas consisting of positively charged ions, negatively charged electrons, and neutral particles co-existing together. Energy is required to extract electrons from a stable atom for generating plasma. Any ionized gas cannot be called as plasma. There should be some degree of ionization. Thus plasma is defined as *a quasi-neutral gas consisting of both charged and neutral particles*.

There are three mechanisms to ionize the gas for plasma generation. These are (i) thermal ionization, (ii) electrical ionization, and (iii) radiant ionization.

In case of thermal ionization, the electron–ion pair generated due to the thermal excitation is unstable. When the temperature and electron density are high enough, the each recombination is followed by ionization and the plasma is able to maintain itself. However, the required temperature for this process is at least 10,000 °C, which is more than what any metal can withstand.

The electrical ionization is done by applying high-intensity electric field to the gas. The electrons get excited and move out of the atoms. These accelerated electrons collide with neutral atoms resulting in further ionization. The storm lightning comes under this type of ionization. The radiant ionization is done through electromagnetic radiation. The incident photons should have energy higher than that of ionization threshold. For example, in ionosphere, ionization takes place due to the UV radiation from the Sun.

It is noted that the plasma should be produced in such a way that it should allow independent variation of plasma parameters. Furthermore, it should not enhance the other types of signatures such as infrared or optical. The basic parameters representing plasma include plasma frequency, density, and Debye length. A brief discussion of these parameters is given below:

Plasma Frequency: A fundamental property of the plasma is its tendency to preserve its macroscopic electrical neutrality. In equilibrium state, the plasma remains uniform and stable. Every elemental volume in plasma has equal number of electrons and ions.

If due to some reason an imbalance of charge occurs, a large electrostatic force is generated resulting in electron plasma oscillations. These oscillations enable the plasma to retain its neutrality on an average over a short period of time (Seshadri 1973).

If an excess negative charge is introduced in the plasma (Fig. 2), a radial electric field is produced due to this electric charge. The electrical field forces the electrons to move radially outward. This gives kinetic energy to the electrons, which moves out of the region resulting in net positive charge. This reverses the direction of electric field, and hence the movement of electrons into the region. This continuous back and forth movement of electrons gives rise to electron plasma oscillations. These are usually very rapid oscillations so that massive ions do not have time to

Fig. 2 Electric field produced by the excess charge

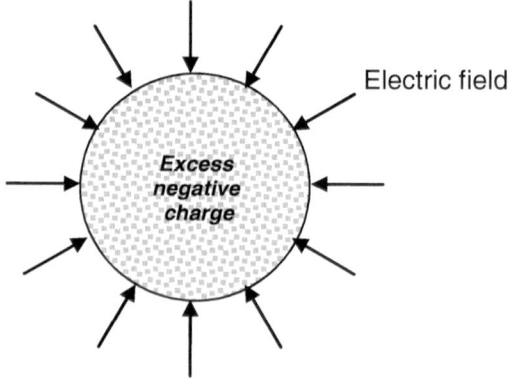

respond to the oscillating field and may be considered as fixed. The frequency of such oscillations is termed as plasma frequency.

The total charge inside the spherical region averaged over a period of these oscillations is zero. This preserves the electrical neutrality on average (Chen 1974). The plasma frequency is expressed as (Ginzburg 1961)

$$\omega_p = \sqrt{\frac{n_o e^2}{m_e \varepsilon_o}} \tag{2}$$

where n_o is the equilibrium electron density, e is the electron charge ($=1.6 \times 10^{-19}$), m_e is the electron mass ($=0.91 \times 10^{-30}$ kg), and ε_o is the permittivity of free space. For example, the plasma frequency for $n_o = 10^{23}$ m^{-3} is 2.8 THz.

Since the plasma oscillation is independent of propagation constant k, it has zero group velocity, i.e., no propagation in infinite system. However, this is not true for a finite system. Here, the plasma oscillations propagate through fringing electric field (Chen 1974). The thermal motion of electrons affects the propagation of plasma oscillations by changing its oscillating frequency by

$$\omega^2 = \omega_p^2 + \frac{3}{2}k^2 v_{th}^2 \tag{3}$$

where $v_{th}^2 = \frac{2KT_e}{m_e}$; v_{th} is the thermal velocity, K is the Boltzmann's constant, T_e is the electron temperature, and k is the propagation vector.

Debye Length: As discussed above, the plasma has tendency to shield the external applied electric field. The plasma oscillations help the plasma to sustain the neutrality in the macroscopic scale. The potential falls away through a characteristic distance, called the Debye length, expressed as (Seshadri 1973)

$$\lambda_D = \frac{1}{\omega_p}\sqrt{\frac{KT_e}{m_e}} \tag{4}$$

A Debye sphere is defined as a sphere with its center coinciding with the location of test particle and radius equal to the order of magnitude of Debye length. The Debye length is the shielding distance or thickness of sheath.

Substituting (2) in (4), the Debye length is expressed as

$$\lambda_D = \sqrt{\frac{KT_e}{4\pi n_o e^2}} \tag{5}$$

The Debye length depends on the density of electrons. For high electron density, λ_D would be small. Furthermore, λ_D is directly proportional to the electron temperature T_e. As an example, for an ionosphere, $n_o \approx 10^{12}$ m^{-3}, $T_e = 1000$ °K, the Debye length is 2×10^{-3}m.

In order to define a term "quasi-neutrality," let us consider a finite system with dimension L, such that $L \gg \lambda_D$. Thus for an external potential applied, the shielding distance would be small as compared to the dimension of the system. In simpler words, the bulk of the plasma would be free of the applied electric potential, and hence the field. One should remember that Debye shielding is valid only for large number of particles within the Debye sphere. This is due to the fact that the electron plasma oscillations are due to the collective behavior of the plasma particles. The number of particles in Debye sphere (Chen 1974) is given by

$$N_D = n_o \frac{4}{3} \pi \lambda_D^3 \tag{6}$$

For collective behavior of plasma, the conditions required to be satisfied are (i) $\lambda_D \ll L$, (ii) $N_D \gg 1$, and (iii) $\omega\tau > 1$ (τ is mean time between the collisions of neutral atoms).

1.2 Mechanisms to Control Plasma Performance

In the absence of magnetic field, the electron density and collision frequency are important parameters that control the performance of the plasma toward RCS reduction (Yu et al. 2003).

Dielectric permittivity: The complex dielectric permittivity of the plasma medium (Ginzburg 1961) is given by

$$\varepsilon_r = 1 - \frac{\omega_p^2}{\omega(\omega - iv)} = 1 - \frac{\omega_p^2}{(\omega^2 + v^2)} - \frac{i\omega_p^2 v}{\omega(\omega^2 + v^2)} \tag{7}$$

Here ω_p is the plasma frequency, ω is the incident wave frequency, and v is the collision frequency.

Since the plasma permittivity depends on the frequency of incident wave, the refractive index of the plasma also changes with the incident wave. For unmagnetized cold collisionless plasma, the refractive index (Ruifeng and Donglin 2003) is expressed as

$$n = \left(1 - \frac{n_e e^2}{\varepsilon_0 m_e \omega^2}\right)^{\frac{1}{2}} = \left(1 - \frac{\omega_{pe}^2}{\omega^2}\right)^{\frac{1}{2}} \tag{8}$$

When $\omega/\omega_{pe} \lesssim 1$, then n will become imaginary and EM wave will not propagate through the plasma. On the other hand, for $\omega/\omega_{pe} \gtrsim 1$, n becomes real, thus EM wave may propagate through the plasma. The former case is similar to frequency band suppression, and latter one represents band pass of a high-pass filter.

Propagation constant: The complex propagation of the medium is expressed as

$$\gamma = jk_{\mathrm{o}}\sqrt{\varepsilon_{\mathrm{r}}} \tag{9}$$

where ε_r is the dielectric permittivity of the plasma, $k_{\mathrm{o}} = \frac{2\pi}{\lambda}$.

When the collision in the plasma is negligible, (9) reduces to

$$\gamma = jk_{\mathrm{o}}\sqrt{1 - \frac{\omega_{\mathrm{p}}^2}{\omega^2}} \tag{10}$$

Electron collision frequency: When there are electron collisions within the plasma, EM wave undergoes attenuation while propagating through the plasma. Hence, the reflected power will decrease when the collision frequency increases. When EM wave enters into the plasma, the collisions within the plasma get enhanced. According to the statistical theory, frequency of the electron collision is expressed as

$$v_{\mathrm{c}} = \frac{4}{3}\pi\alpha^2 N v_{\mathrm{av}} \tag{11}$$

where α is the diameter of the neutral particles, N is the number of density, and v_{av} is the average velocity of electron.

Effect of plasma frequency: There are three cases of propagation depending on the plasma frequency (Table 1). One can infer that if the EM frequency is less than the plasma frequency, the plasma behaves as a good reflector to the incident wave. Otherwise, the incident wave gets attenuated by the factor $e^{-\alpha z}$.

In other words, if the EM wave enters into the plasma medium, the plasma acts as a good absorber. The plasma density which determines the plasma frequency also controls the cut-off and the absorption within the plasma.

The above discussion is valid only till there is no collision in the plasma medium. If collisions happen, then the plasma performance will change altogether (Jenn 2005). The dielectric constant of plasma and the loss that an incident EM wave (1 GHz) undergoes within the plasma for different collision frequencies are given in Table 2. The absorption characteristics of plasma depend on the collision frequency and incident wave frequency (Gregoire et al. 1992). The effect of

Table 1 Effect of plasma frequency (ω_{p}) on plasma performance

Incident wave frequency (ω)	Propagation constant (γ)	Effect
$\omega < \omega_{\mathrm{p}}$	Real	Wave propagates ($e^{-j\beta z}$)
$\omega > \omega_{\mathrm{p}}$	Imaginary	Wave attenuates ($e^{-\alpha z}$)
$\omega = \omega_{\mathrm{p}}$	Zero	Boundary between propagation and attenuation

Table 2 Dielectric constant and loss for different collision frequencies

Collision frequency v_c (per second)	Electron density n_e (per m^3)	Plasma frequency ω_p (MHz)	Dielectric constant of plasma	Loss (dB/m)
0	10^{15}	284	0.9196−j0	0
	10^{16}	897	0.196−j0	0
10^7	10^{15}	284	0.9196−j0.0001	2
	10^{16}	897	0.196−j0.0013	0.263
10^9	10^{15}	284	0.9196−j0.01251	3
	10^{16}	897	0.2159−j0.1248	23.5

Table 3 Effect of collision frequency (v) on plasma performance

Incident wave frequency (ω)	Absorption characteristics	Reason
$\omega \ll v$	Poor absorption	Electrons acquire little energy before colliding with neutrals
$\omega \sim v$	Peak absorption	Maximum energy transfer
$\omega \gg v$	Poor absorption	Due to the low collision rate electrons simply oscillate in the EM wave field

collision frequency with respect to the incident wave frequency on the plasma performance is shown in Table 3.

Effect of electron temperature: During collisions, the electron momentum transfer rate is a function of electron temperature and gas species (Vidmar 1990). For example, at a pressure of 760 torr and 1000 K, the thermalizations in He and air require 50 ns and 20 ns, respectively. If thermalization is faster than the life time of plasma (τ), the electron temperature is taken to same as the ambient temperature. Plasma life time (τ) is the time required by a plasma with initial density n_o to reduce in concentration by a factor of $1/e$.

The plasma life time depends on electron density and pressure. It increases as the pressure decreases and the electron density decreases. On the other hand, if thermalization is slower, then a higher temperature is taken for the electron temperature. Helium (He) does not form negative ions, which minimize the main electron attachment mechanism, i.e., longer life time is possible with He. Since high purity He is expensive, trace amount of air in He can be used for plasma generation. The N_2 and O_2 impurities will rapidly deionize the He plasma.

Williams and Geotis (1989) studied the effect of lightning discharge on radar waves. Plasma is formed as a result of ionization of atmosphere by lightning. This plasma affects the propagation of radar waves. It is reported that the plasma frequency is determined by the electron density, which depends on electron temperature. The plasma frequency rises very rapidly with the temperature in the 2000–4000 K. The rate of increase, however, becomes slower at higher temperatures.

For temperatures greater than 5000 K, the plasma frequency is greater than all radar frequencies. Thus in such conditions, the plasma acts as good reflector. It is to be noted that in hot air, plasma collisions are very likely to exist. The higher the electron temperature, the more will be the collision frequency. This makes the plasma highly reflective, making it less preferred for stealth application.

Effect of magnetic field: If a static uniform magnetic field is applied, the plasma medium becomes anisotropic. Considering non-collision case, the non-zero elements of the permittivity matrix (Ginzburg 1961), when the applied magnetic field (B_o) is in z-direction, are expressed as

$$\varepsilon_{xx} = \varepsilon_{yy} = \varepsilon_o \left(1 + \frac{\omega_p^2}{\omega_c^2 - \omega^2} \right) \tag{12a}$$

$$\varepsilon_{xy} = \varepsilon_{yx}^* = \frac{j\omega_p^2 \left(\frac{\omega_c}{\omega} \right) \varepsilon_o}{\omega_c^2 - \omega^2} \tag{12b}$$

$$\varepsilon_{zz} = \varepsilon_o \left(1 - \frac{\omega_p^2}{\omega^2} \right) \tag{12c}$$

where ω_c is the cyclotron frequency, given by

$$\omega_c = \frac{-eB_o}{m_e} \tag{13}$$

A moving electron in the applied magnetic field will rotate with cyclotron frequency even in the absence of the incident EM wave. If the incident wave is synchronized with the electron motion (i.e., $\omega = \omega_c$), the electron can achieve higher velocities, extracting energy from the incident wave. The applied magnetic field, which determines the cyclotron frequency, provides another option to control the plasma parameters.

If collisions are considered in the plasma, the permittivity matrix in the presence of external magnetic field is modified by the collision frequency. It is expressed as

$$\varepsilon_{ij}(\omega) = \begin{vmatrix} \varepsilon_{xx}(\omega) & -j\varepsilon_{xy}(\omega) & 0 \\ -j\varepsilon_{yx}(\omega) & \varepsilon_{xx}(\omega) & 0 \\ 0 & 0 & \varepsilon_{zz}(\omega) \end{vmatrix} \tag{14}$$

$$\text{where } \varepsilon_{xx}(\omega) = \varepsilon_{yy}(\omega) = \varepsilon_o \left(1 - \frac{\left(\frac{\omega_p}{\omega} \right)^2 \left(1 - j\frac{v}{\omega} \right)}{\left(1 - j\frac{v}{\omega} \right)^2 - \left(\frac{\omega_c}{\omega} \right)^2} \right) \tag{15a}$$

Table 4 Effect of applied magnetic field on the plasma performance

Type of propagation	Effect on performance
Perpendicular polarization ($k \perp B_o$)	$E \parallel B_o$: Wave unaffected by B_o (Ordinary wave)
	$E \perp B_o$: Effective refractive index and phase velocity affected by applied magnetic field (Extra ordinary wave)
Parallel polarization ($k \parallel B_o$)	Two circularly polarized components of wave will travel in different velocities, and hence plane of polarization will rotate with distance (Faraday rotation)

$$\varepsilon_{xy}(\omega) = \varepsilon_{yx}(\omega) = \varepsilon_0 \frac{\left(\frac{\omega_p}{\omega}\right)^2 \left(\frac{\omega_c}{\omega}\right)}{\left(1 - j\frac{v}{\omega}\right)^2 - \left(\frac{\omega_c}{\omega}\right)^2} \tag{15b}$$

$$\varepsilon_{zz}(\omega) = \varepsilon_0 \left(1 - \frac{\omega_p^2}{\omega(\omega - jv)}\right) \tag{15c}$$

The magnetic field parallel to z-axis is the simplest case, and the permittivity matrix will be more complex if the applied magnetic field is inclined to z-axis. EM propagation characteristics of plane wave in plasma can be explained on the basis of the directions of applied magnetic field and the incident wave propagation. Table 4 describes the effect of applied magnetic field on the performance of plasma.

1.3   Plasma Stealth: History and Future

Russian aircraft industry has developed plasma-based stealth technology. Almost decade has passed, since Russian officials had accepted that they are capable of using plasma screens to cushion and disperse radar waves. This plasma screen technology was reported to be applicable to any aerospace vehicle and ships. The technology is based on the principle other than used in U.S-based stealth aircraft (e.g., F-117, B-2).

The Ekurdsh Scientific Research Center had introduced the third-generation stealth system, which uses plasma around the exterior of aircraft without altering the external contours of the aircraft (Beskar 2004). It is anticipated that the plasma stealth has been incorporated in various versions of Su-7 and MiG-35, Russian fighter aircraft. There have also been claims that the Russians have achieved RCS reduction by two orders of magnitude through plasma stealth device.

Plasma Stealth in Future: One of the main contributions to the aircraft RCS is from the antenna mounted on the surface of the aircraft. Plasma can be used for antenna surfaces to generate low-observability characteristics (Alexef et al. 1998). The hollow glass tube filled with low-temperature plasma acts as a plasma-based antenna. Such antenna can make the surface completely radar transparent when not in use as radiating element (Cadirci 2009), (Anderson et al. 2006). Such plasma-based antenna can appear and disappear in a few microseconds.

One of the criticisms to the plasma-based antennas is that they are fragile. The robustness of such antennas can be enhanced by embedding it in an epoxy block (Anderson et al. 2006). Furthermore, plasma can be used as a substitute for the metallic elements in the frequency selective surfaces (FSS). The plasma-based FSS can be tuned by varying the plasma density. This offers improved shielding along with reconfigurability and stealth.

The antenna dimension, plasma frequency, and collision frequency control the antenna gain, input impedance, efficiency, and the RCS (Sadeghikia and Kashani 2013). In order to achieve RCS reduction, large collision frequency is preferred. As the collision frequency increases the efficiency, peak input impedance and the gain of the antenna are decreased compared to the conventional metallic antenna. However, the radiation pattern remains unaffected by the increase in the collision frequency.

At higher plasma frequency w.r.t. the excitation frequency, the values of gain, radiation pattern, and RCS are close to that of a metallic antenna. The variation in plasma frequency affects the input impedance of the antenna. However, the detailed RCS characteristics of plasma antennas are not available. Furthermore, the deployment of plasma antenna to reduce RCS is still under research.

2 EM Wave Propagation Through Plasma

When a plane wave propagates through plasma, it undergoes absorption, reflection, and transmission. A plane wave in plasma, like all other lossy medium, obeys Maxwell's equations:

$$\nabla \times E = -j\omega\mu_r\mu_o H \tag{16a}$$

$$\nabla \times H = (\sigma + j\omega\varepsilon_r\varepsilon_o)E \tag{16b}$$

where the complex dielectric permittivity $\bar{\varepsilon_r}$ is given by

$$\bar{\varepsilon}_r = \varepsilon_r - j\frac{\sigma}{\omega\varepsilon_o} \tag{17}$$

Thus (16b) becomes

$$\nabla \times H = j\omega\bar{\varepsilon}_r\varepsilon_o E \tag{18}$$

The resulting wave equation is expressed as

$$\nabla^2 E = \frac{\mu_r\bar{\varepsilon}_r}{c^2}\frac{\partial^2 E}{\partial t^2} = -\omega^2\frac{\mu_r\bar{\varepsilon}_r}{c^2}E = \bar{\gamma}E \tag{19}$$

where c is the speed of light and $\bar{\gamma}$ is the complex propagation constant, given by

$$\bar{\gamma} = j\frac{\omega}{c}\sqrt{\mu_r\bar{\varepsilon}_r} \tag{20}$$

2.1   Wave Propagation in Magnetized Plasma

The interaction of EM waves with magnetized plasma is of interest, especially in the region near the electron cyclotron frequency. This is because the magnetized plasma shows large absorption (up to 40 dB) near the electron cyclotron frequency (Roth 1994). This feature can be exploited in plasma-based cloaking of space craft and earth satellite, where the environment is suitable to maintain the magnetized plasma. On the other hand in case of aircraft, the environment around is atmospheric air, making generation and maintenance of plasma a challenge.

2.1.1   Stationary Plasma Slab

Considering non-uniform plasma slab (Fig. 3a), the absorption, reflection, and transmission in cold weakly ionized collisional magnetized plasma can be analyzed (Laroussi and Roth 1993). Non-uniform plasma slab can be considered a stack of sub-slabs (Fig. 4). In each sub-slab, plasma density is constant and overall plasma density across the slab follows a parabola (Fig. 3b). The complex dielectric constant for plane wave through the cold plasma at arbitrary angle of incidence (Froula et al. 2011) is given by

$$\bar{\varepsilon}_r = 1 - \frac{\frac{\omega_p^2}{\omega^2}}{\left[1 - j\frac{\nu}{\omega} - \frac{\frac{\omega_c^2}{\omega^2}\sin^2\theta}{2\left(1 - \frac{\omega_p^2}{\omega^2} - j\frac{\nu}{\omega}\right)}\right] \pm \left[\frac{\frac{\omega_c^4}{\omega^4}\sin^4\theta}{4\left(1 - \frac{\omega_p^2}{\omega^2} - j\frac{\nu}{\omega}\right)^2} + \frac{\omega_c^2}{\omega^2}\cos^2\theta\right]^{\frac{1}{2}}} \tag{21}$$

Fig. 3   a Microwaves incident on non-uniform plasma. **b** Parabolic density profile of plasma

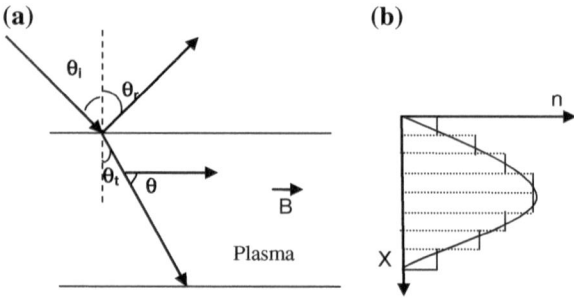

Fig. 4 Reflection and refraction from successive plasma slabs

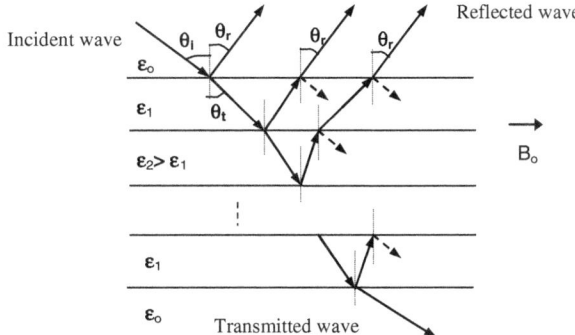

where ω_p is the plasma frequency, ν is the collision frequency, ω_c is the gyrofrequency of electron, and θ is the angle of propagation w.r.t. the magnetic field. The $-$ sign represents the right-hand polarization and $+$ for the left-hand polarization.

The reflection from each slab boundary is calculated based on the condition of the continuity of the wave impedance at the slab boundary. The reflection coefficient (Born and Wolf 2002) is given by

$$\Gamma(\omega, X_{i+1}) = \frac{\frac{\bar{\varepsilon}_r(\omega, X_{i+1})}{\bar{\varepsilon}_r(\omega, X_i)} \cos\theta_i - \left(\frac{\bar{\varepsilon}_r(\omega, X_{i+1})}{\bar{\varepsilon}_r(\omega, X_i)} - \sin^2\theta_i\right)^{\frac{1}{2}}}{\frac{\bar{\varepsilon}_r(\omega, X_{i+1})}{\bar{\varepsilon}_r(\omega, X_i)} \cos\theta_i + \left(\frac{\bar{\varepsilon}_r(\omega, X_{i+1})}{\bar{\varepsilon}_r(\omega, X_i)} - \sin^2\theta_i\right)^{\frac{1}{2}}} \tag{22}$$

where θ_i is the incident angle of the EM wave and X_i is the location of ith slab. The total reflection coefficient $\Gamma_T(\omega)$ is the sum of partial reflections from the slab boundary, weighted by the respective attenuation factor (Laroussi and Roth 1993). It is given by

$$\Gamma_T(\omega) = \sum_{i=1}^{N} \Gamma(\omega, X_i)(q(\omega, X_i))^2,$$

$$q(\omega, X_i) = 1 - \exp\left(-\alpha(\omega, X_i) \cdot \frac{X_i}{\sin\theta}\right) \tag{23}$$

where $\alpha(\omega, X_i)$ is the attenuation coefficient at respective plasma slab. The total reflected and transmitted power are expressed as

$$P_R = P_{\text{inc}}|\Gamma_T(\omega)|^2 \tag{24}$$

$$P_T = P_{\text{inc}} \prod_i F_i \tag{25a}$$

where p_{inc} is the incident power and F_i is expressed as

$$\prod_i F_i = \exp\left(\frac{-\alpha_i \cdot d}{\sin \theta}\right) \tag{25b}$$

where α_i and d are the attenuation coefficient at ith slab and slab width, respectively. Thus the power absorbed can be obtained as

$$P_{abs} = P_{inc} - P_R - P_T \tag{26}$$

The scattering matrix method (SMM) (Hu et al. 1999) can be used to analyze the reflection, absorption, and transmission through non-uniform magnetized plasma. The SMM calculates the partial reflection, absorption, and transmission in each layer using backward recursion formula in terms of 2×2 S-parameter matrices. The method provides solution for both exponential and parabolic density profiles for normal incidence case.

Table 5 describes the trend of reflected, transmitted, and absorbed power when the incident wave is normal to the applied magnetic field (Hu et al. 1999). The oblique incidence case (60°) is tabulated in Table 6 (Laroussi and Roth 1993), using Fresnel's reflection coefficients. The plasma density profile is taken parabolic. The collision frequency considered is of MHz range, while the incident frequency (ω), gyrofrequency (ω_c), and plasma frequency (ω_p) are in GHz range.

When $v \ll \omega$ and $\omega_p \ll \omega_c$, frequency (ω_{max}) at which peak of reflection, absorption, and transmission occurs can be obtained from the imaginary part of the propagation constant, i.e.,

$$\text{Im}(\varepsilon_r) = -\frac{v\omega_p^2}{\omega^3\left(1 - \frac{\omega_c^2}{\omega^2}\left(1 + \frac{\omega_p^2}{\omega^2}\right)\right)^2 + v^2\omega} \tag{27a}$$

Table 5 Performance analysis of plasma slab (Normal incidence)

Plasma parameter	Reflected power	Absorbed power	Transmitted power
increasing peak electron density ($\omega_p < \omega$)	Increases	Increases	
	Bandwidth remains same, peak shifts to higher frequency		
Increasing v (in MHz) ($v < \omega$, ω_p)	Decreases	Decreases	Increases
	Bandwidth and f_{max} remains the same		
Electron density profile	Parabolic profile has high value than exponential profile.		Exponential profile has high value than exponential profile.

Table 6 Performance analysis of plasma slab (Oblique incidence to magnetic field, 60°)

Plasma parameter	Reflected power	Absorbed power	Transmitted power
Increasing peak electron density ($\omega_p < \omega$)	Increases	Increases	Decreases
	Bandwidth increases, peak shifts to higher frequency		
Increasing ν (in MHz) ($\nu < \omega, \omega_p$)	Decreases Bandwidth and f_{max} remains the same	Decreases	Increases
		Slight increase in bandwidth, f_{max} remain the same	
Small angle of incidence	High for smaller angles, f_{max} shifts slightly to higher frequency and bandwidth slightly increases due to shift in lower frequency limit	High	Low
		f_{max} remain the same, bandwidth increases due to shift in lower frequency limit	

For maximum value of Im(ε_r), the first term of denominator should be zero, i.e.,

$$1 - \frac{\omega_c^2}{\omega^2}\left(1 + \frac{\omega_p^2}{\omega^2}\right) = 0 \tag{27b}$$

This implies that

$$\omega_{max} = \sqrt{\frac{\omega_c^2 + \sqrt{\omega_c^4 + 4\omega_c^2\omega_p^2}}{2}} \approx \omega_c + \Delta\omega_c \tag{27c}$$

i.e., ω_{max} increases with increase in electron cyclotron frequency and electron density (ω_p is function of electron density from (2)). This shows that for $\omega_p \ll \omega_c$, the peak will be close to cyclotron frequency (ω_c). There is an optimum cyclotron frequency at which reflection and absorption attain maximum value.

Using (27c), (27a) reduces to

$$Im(\varepsilon_r) = -\frac{\omega_p^2}{\nu\omega_{max}} \tag{28}$$

It may be inferred that the loss and, hence, attenuation constant are directly proportional to the electron density while inversely proportional to the collision frequency (Hu et al. 1999).

Next is the case when the wave is obliquely incident to the non-uniform magnetized plasma having linear electron density profile (Mo and Yuan 2008). Table 7 presents the absorption behavior of plasma slab for different plasma parameters.

If the plasma density profile is partially linear and sinusoidal plasma density profile (Fig. 5b), EM propagation shows different trends (Gruel and Oncu 2009). The plasma medium is taken as cold, steady state, collisional, and weakly ionized.

Table 7 Performance analysis of plasma slab (Oblique incidence to the plasma slab, 30°)

Plasma parameter	Absorbed power
High peak electron density	Broader absorption band
High v (in GHz) ($v > \omega_p$, ω)	Deeper and increased absorption band
Large incident angle	Broader absorption band
High magnetic field strength	Band moves to high-frequency region

Fig. 5 **a** Electromagnetic wave propagation through the plasma layer. **b** Electron density profile through the plasma

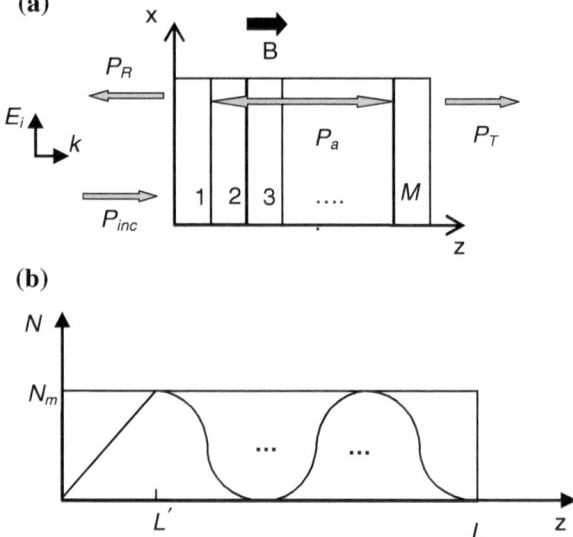

The magnetic field applied is uniform and parallel to the magnetized plasma slab (Fig. 5a). The first part of the plasma has linearly varying density profile, in order to have better match of the EM wave with the plasma slab. The linear profile is followed by sinusoidal density profile in order to obtain improved absorption and transmission performances. For a normally incident plane wave with magnetic field parallel to propagation direction ($\theta = 0°$) (Fig. 5a), the complex dielectric constant can be determined from (21):

$$\bar{\varepsilon}_r = 1 - \frac{\frac{\omega_p^2}{\omega^2}}{1 - j\frac{(v-\omega_c)}{\omega}} \tag{29}$$

Again, the plasma layer is divided into equal-width thin sub-layers, with uniform plasma in each sub layer (Fig. 5a). The electron density profile N along the plasma layer is given by (Gruel and Oncu 2009)

$$N = \begin{cases} \frac{N_m z}{L'} & z < L' \\ N_m \left(0.5 + 0.5 \cos\left(\frac{\pi(z-L')}{(L-L')}\right)\right) & z > L' \end{cases} \tag{30}$$

where N_m is the maximum electron density, L is the thickness of plasma layer, and L' is length of linear profile.

The refection coefficient at the $(i + 1)$th interface for normal incidence $(\theta = 0°)$ can be obtained from (22):

$$\Gamma(i+1) = \frac{\sqrt{\bar{\varepsilon}_r(i+1)} - \sqrt{\bar{\varepsilon}_r(i)}}{\sqrt{\bar{\varepsilon}_r(i+1)} + \sqrt{\bar{\varepsilon}_r(i)}} \tag{31}$$

As the wave propagates through the plasma layer, the transmitted wave power exponentially reduces. The total reflected power from M layered plasma is given by

$$P_R = P_{inc}\left\{|\Gamma(i)|^2 + \sum_{j=2}^{M}\left(|\Gamma(j)|^2 \prod_{i=1}^{j-1} \exp(-4\alpha(i)d)\left(1 - |\Gamma(i)|^2\right)\right)\right\} \tag{32}$$

where d is the thickness of the sub-slab, $\alpha(i)$ is the attenuation coefficient at the ith slab, and P_i is the incident power. The multiple reflections at the interface are neglected.

The total transmitted power is given by

$$P_T = P_{inc} \prod_{i=1}^{M} \exp(-2\alpha(i)d)\left(1 - |\Gamma(i)|^2\right) \tag{33}$$

and the absorbed power is expressed as

$$P_a = P_{inc} - (P_R + P_T) \tag{34}$$

The plasma performance with such electron density profile is summarized as Table 8.

Table 8 Performance analysis of plasma slab (Fig. 5)

Plasma parameter	Plasma performance
Increasing electron density	Absorbed power increases (f_{max} remains the same, band increases)
Increasing v (in GHz)	Widens the absorption band (f_{max} remains the same)
Increasing B_o	Peak absorption and zero transmission near ω_c
	Absorption band shifts to high-frequency region (bandwidth remains the same)
Increasing $\frac{L'}{(L-L')}$ ratio	Reflected power decreases
	Absorption band is widened (f_{max} remains the same)

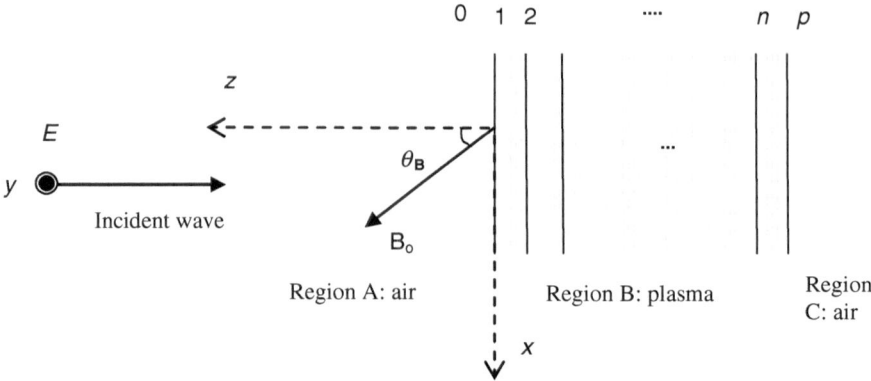

Fig. 6 Non-uniform magnetized plasma slab model

Recently, *Propagator Matrix Method* (PMM) is proposed by Yin et al. (Yin et al. 2013) for analyzing the EM wave interaction with plasma slab. Non-uniform magnetized plasma with arbitrary magnetic declination angle and bi-exponential density profile is considered (Fig. 6). It is shown that the plasma exhibits transformation in polarization when a plane wave is normally incident. This polarization transformation depends on the plasma parameters.

The bi-exponential distribution profile can be expressed as

$$n_e(z) = \begin{cases} n_p \exp(-\frac{z+L_1}{z_{10}}) & -L_1 \leq z \leq 0 \\ n_p \exp(\frac{z+L_1}{z_{20}}) & -L_2 \leq z \leq L_1 \end{cases} \quad (35)$$

where z is the radial distance, n_p is the peak electron density, z_{10} and z_{20} represent the curve shapes, L_1 is the location where the electron density is maximum, and L_2 is the plasma thickness.

The propagator matrix (4×4) in the plasma region is expressed as the product of the propagator matrix in each layer. It represents the reflected wave at the 0th layer and the transmitted wave beyond pth layer.

The normalized absorbed power through the plasma slab can be obtained from the total reflected and transmitted powers:

$$P_a = 1 - P_{co}^r - P_{cross}^r - P_{co}^t - P_{cross}^t \quad (36)$$

where P_{co}^r and P_{cross}^r are the co- and cross-polarized reflected powers, and P_{co}^t and P_{cross}^t are the co- and cross-polarized transmitted powers, respectively.

For unmagnetized plasma slab, the cross-polarized components of reflected and transmitted power will vanish due to the isotropic nature. The propagation and polarization characteristics of the plasma depend upon electron density, collision frequency, cyclotron frequency, and magnetic declination angle.

In general, $P_{cross}^t > P_{cross}^r$ for magnetized plasma slab. Moreover, the plasma frequency varies with the plasma thickness for bi-exponential density profile. This variance of plasma frequency is greater for higher electron density (Yin et al. 2013). The dependence of reflected, transmitted, and absorbed power of plasma slab with b-exponential density profile on various plasma parameters is summarized in Table 9.

Table 9 Performance analysis of plasma slab with bi-exponential density profile and inclined magnetic field

Power	Condition	Increase in ω	Increase in n_p	Increase in ν	Increase in B_o	Decrease in θ_B
P_{co}^r; P_{cross}^r		Minimum at $\omega = \omega_c$	Regular variation		Irregular variation	
	$\omega < \omega_c$	Decreases to minimum at ω_c	Does not have any effect	P_{co}^r decreases Not much variation in P_{cross}^r	Small Peak appears before reaching ω_c (Peak value increases with B_o)	
	$\omega > \omega_c$	P_{co}^r varies with a peak value P_{cross}^r varies with two peaks	P_{co}^r peak and bandwidth increases P_{cross}^r: 2nd peak and its frequency increases; first peak remains same	Peak value of both $P_{co}^r P_{cross}^r$ decreases	Fluctuation increases	P_{co}^r peak value and the bandwidth decreases (except at 0°) P_{cross}^r vanishes at 90°
P_{co}^t		Shows turning point (ω_t)	ω_t increases	No effect	ω_t increases	Slight increase in ω_t
	$\omega < \omega_t$	Varies slowly with a small peak	Frequency at which small peak occurs increases but with same peak value	Peak value decreases	Peak value and the corresponding frequency increases (except at 90°, peak vanishes)	
	$\omega > \omega_t$	Increases fast to approach value of 1	Faster rate of increase			

(continued)

Table 9 (continued)

Power	Condition	Increase in ω	Increase in n_p	Increase in v	Increase in B_o	Decrease in θ_B
P^t_{cross}		Shows an optimum frequency (ω_{opt}) at which peak occurs	Peak value and ω_{opt} increases (bandwidth increases)	Peak value decreases; same ω_{opt} (bandwidth increases)	Peak, ω_{opt} increases (P^t_{cross} vanishes at 90°)	
P_a		Increases, attains maximum (at ω_c) and then decreases	Band width increases No change in peak		Peak value decreases Band width increases	No change (except at 0°, both peak and bandwidth are less)

2.1.2 Moving Plasma Slab

If the plasma slab is in motion (Fig. 7), the propagation characteristics of EM wave with the slab will be different due to relativistic effect. Power reflection and transmission coefficients of an normally incident EM wave on a moving plasma slab with uniform velocity (v) can be determined by imposing boundary conditions in the primed frame, followed by the relativistic transformations of plasma parameters from primed to un-primed frame (Chawla and Unz 1969). Due to motion of plasma slab, the reflected frequency and the wave number suffer Doppler effect, whereas the transmitted frequency and the wave number remain unaffected. Moreover, the Doppler effect is independent of the slab medium.

Table 10 summarizes the propagation characteristics for a moving magnetized plasma slab. It is to be noted that in Fig. 7, the magnetized plasma is moving perpendicular to the interface.

However, if the plasma slab moves with uniform velocity in a direction parallel to the interface (Fig. 8), both the reflected and transmitted waves will not suffer from Doppler shift (Singh and Shekhawat 1983). The external magnetic field (B_o) is obliquely applied to the plasma. The effect of plasma parameters on the propagation phenomena is summarized in Tables 11 and 12. Both reflection and transmission

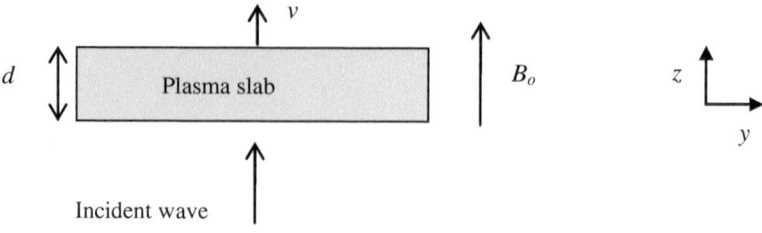

Fig. 7 Magnetized plasma slab moving perpendicular to the interface

Table 10 Performance analysis of moving magnetized plasma

Power coefficients	Comments	
Reflection (R)	For negative β, R shows some oscillations	
	R decreases as β increases	
	$\beta = v/c$	ω_c/ω
	$\beta < 0: R > 1$	R_{max} increases as ratio ω_c/ω increases
	$\beta > 0: R < 1$	
Transmission (T)	$T < 1$ for all cases	
	$\omega_c/\omega = 0$: T tends to zero as β increases	
	$\omega_c/\omega \neq 0$: T tends to value of 1 as β increases	
	Ripples arises and increases as the strength of applied magnetic field increases	
$(R + T)$	$\beta < 0: R + T > 1$	
	$\beta = 0: R + T = 1$	
	$\beta > 0: R + T < 1$	

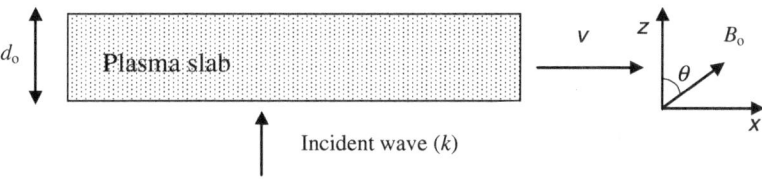

Fig. 8 Plasma slab moving parallel to the interface and with obliquely applied magnetic field

Table 11 Variation in coefficients w.r.t. plasma slab velocity ($\beta = v/c$)

Plasma slab vel. (β)	$\omega_p/\omega < 1$			$\omega_p/\omega > 1$		
	Absorption (A)	Reflection (R)	Trans. (T)	Reflection (R)	Trans. (T)	Absorption (A)
$\beta < 0$	Increases[b]	Slightly increases	Negligible variation	Maximum R and minimum T occurs at particular β value	Negligible variation	
			But shows oscillations at higher θ	T quickly increases till the saturation as β increases		
				R remains maximum for smaller value of β and decreases with increase in β		
$\beta > 0$	Decreases[b]	Shows maximum R and minimum T at particular β^a		Decreases	Slightly increases	Slightly decreases

[a]Peak R, minimum T, and β value changes according to θ values
[b]Variation in A decreases as θ increases for both $\beta < 0$ and $\beta > 0$

Table 12 Variation in coefficients of moving plasma w.r.t. magnetic inclination angle (θ)

$\omega_p/\omega < 1$	$\omega_p/\omega > 1$
Reflection decreases as θ increases	Reflection increases to maximum as θ increases
Transmission decreases to minimum, remains minimum for certain angular range and then increases to maximum as θ increases	Transmission decreases to minimum as θ increases
Absorption increases to maximum, remains maximum for the angles at which minimum transmissions occurs and then decreases as θ increases	Absorption shows random behavior. Absorption and transmission coefficients show minima when reflection coefficient is maximum

coefficients w.r.t. β show different trends for $\omega_p/\omega < 1$ and $\omega_p/\omega > 1$. For $\omega_p/\omega < 1$, transmission minima and reflection maxima occur at positive β value. Moreover, transmission coefficient has higher value as compared to absorption and reflection ones. For $\omega_p/\omega > 1$, transmission minima and reflection maxima occurs at negative β value.

Absorption coefficient has higher value than transmission and reflection ones. The variation in absorption coefficient w.r.t. β shows similar trend for $\omega_p/\omega < 1$ and $\omega_p/\omega > 1$, but the values of absorption coefficient corresponding to $\omega_p/\omega > 1$ are higher than $\omega_p/\omega < 1$.

2.2 Wave Propagation in Unmagnetized Plasma

When an EM wave propagates through unmagnetized plasma, the overall reflection includes reflections at the interface, partial reflections from the bulk plasma, and reflected waves due to collisions. Partial reflections depend on the density gradients, and hence need definition of a density profile. At low-frequency end, reflections are due to density gradient of plasma, while at high frequencies, this is not the case. Rather reflections are due to the decrease in absorption of wave in plasma (Gregoire et al. 1992).

For collisionless case, an incident EM wave propagates through the plasma without attenuation if the incident frequency is greater than the plasma frequency. This is not the case if the radiation frequency is less than the frequency of plasma, i.e., the incident wave undergoes reflections. However, for collisional plasma, incident EM wave gets attenuated instead of undergoing reflections. This is because of the modified complex propagation constant due to the collisions within the plasma.

The electron temperature plays an important role in determining the real and imaginary parts of the complex permittivity. Permittivity of plasma at high temperature is expressed as (Yuan et al. 2010)

$$\varepsilon_2 = \frac{\left[1 - \frac{\omega_p^2}{\omega(\omega - iv)}\right]}{\left[1 + \frac{\omega\omega_p^2 KT_e}{(\omega - iv)^3 m_e c^2}\right]} \tag{37}$$

where K is the Boltzmann's constant and T_e is the electron temperature. The second term in the denominator of above equation may be ignored when the electron temperature is only several electron volts (cold plasma).

The electron temperature affects the collision frequency significantly, i.e.,

$$v = N_o \sigma_e \sqrt{\frac{KT_e}{m_e}} \tag{38}$$

where σ_e is the cross section of the electron elastic collision and N_o is the density of the gas molecules.

The absorption by the plasma is due to the electron neutral collision and due to the cavity resonance effect between the boundaries of the plasma. For high frequency, cavity resonance effect dominates over the collisional absorption. Moreover, the absorption is directly proportional to the plasma thickness. The propagation characteristics of unmagnetized bounded plasma are summarized in Table 13.

An EM wave incident on the bounded plasma rapidly gets absorbed. The field oscillations in the plasma are more prominent for small plasma thickness (Yuan et al. 2010). This is due to the constructive interference of the multiple reflected waves. On the other hand, for large thickness ($d > \pi/k$), the effect of multiple reflections on the field profile within the plasma becomes less.

Table 13 Propagation characteristics of unmagnetized plasma at normal incidence

Parameter	Characteristics
Plasma thickness	• Optimum plasma thickness for minimum reflection • Sheath formed at the boundary of bounded plasma decreases the reflected power without changing the position of the absorption peak and the effective bandwidth
Incident frequency	• Reflected power decreases as the incident frequency increases • At higher frequency, the dependence of reflection on plasma thickness becomes prominent • For high frequencies, reflection increases with the thickness
Collision frequency	Reflection decreases as collision frequency increases
ω_p/ω	For higher ω_p/ω, reflection is more, and independent of plasma thickness

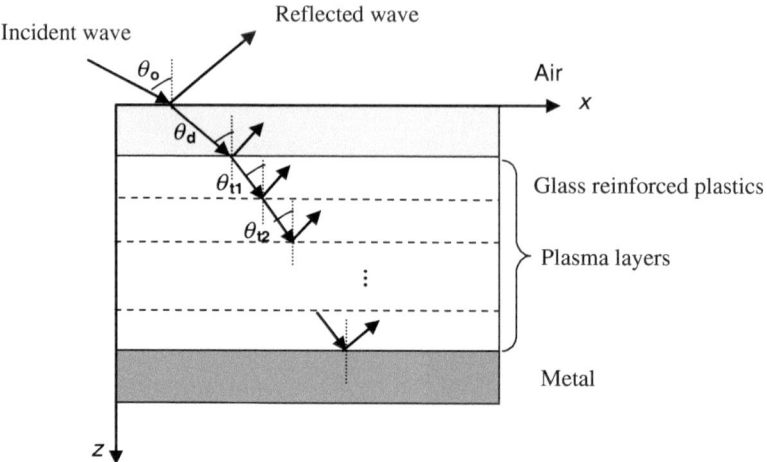

Fig. 9 Plasma with obliquely incident EM wave

Thus in order to have significant absorption, the plasma slab thickness must satisfy the following condition:

$$k = \frac{n\pi}{d} \tag{39}$$

where k is the propagation constant and d is the plasma thickness, $n = 1, 2, 3, \ldots$.

It may be inferred that once the operating frequency of the enemy radar source is known, different plasma parameters can be modulated to achieve plasma stealth.

Let us now consider an EM wave obliquely incident on closed unmagnetized collisional non-uniform plasma (Ma et al. 2008). The plasma layer is divided into n layers (Fig. 9). Each layer has uniform distinct density. Table 14 summarizes the EM wave propagation within unmagnetized bounded plasma, when EM wave is obliquely incident on it.

Cylindrical plasma: The refraction-based characteristics of plasma covering a cylinder have stealth applications in aerospace structures. Ma et al. (2010a) considered a cylindrical perfect conductor with radius r_o covered with concentric plasma envelope, as shown in Fig. 10.

There can be two cases for EM wave incidence. (i) incident EM rays at larger distance from the center ($r_d > r_o$): Such waves have larger angle of incidence and will be refracted by the plasma before they reach the cylindrical conductor; (ii) EM rays at shorter distance from the center ($r_d < r_o$): Such wave may be incident on the cylindrical conductor, and hence reflected. This reflected EM wave may be refracted again by the plasma envelope.

The plasma density is assumed to be function of radius, $n(R) = \frac{R^m}{R_o^m}$, $r_o < R < R_o$. Here, r_o is the radius of the cylindrical conductor, R_o is the radius of the plasma envelope, and r_d is the distance of the EM waves from the center of the conductor.

Table 14 Propagation characteristics of unmagnetized plasma with oblique incidence

Parameter	Characteristics
Incident wave frequency	On increasing incident frequency
	Total reflection occurs at higher angles (Both TE and TM case)
	Reflection loss and bandwidth increases (Both TE and TM case)
	For TE case, peak of reflection loss shifts to higher angle
Incident angle	For both TE and TM case, total reflection takes place at same angle
	TE case: reflection loss shows peak
	TM case: No peak of reflection loss
Temperature	For both TE and TM case, reflection loss increases with temperature
Electron density profile	Exponential profile has wider absorption band compared to parabolic profile

Fig. 10 EM wave incidence on a cylindrical plasma envelope

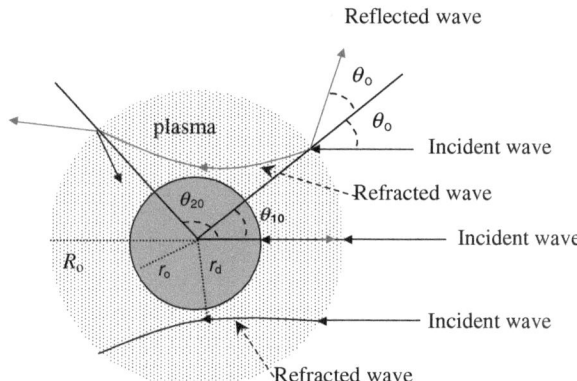

θ_{10} is the angle of incidence of EM wave w.r.t. center of the conductor, θ_o is the angle of incidence of EM wave w.r.t. the surface of plasma, and θ_{20} is the angle of emergence of EM wave w.r.t. center of the conductor. It can be noted that the refraction deviation angle depends upon the angle of incidence.

In other words, the larger the angle of incidence, the greater will be the refraction deviation angle, and hence the lesser will be the backscattering, thereby contributing toward stealth effect.

Moreover, the refraction deviation angle depends on plasma density distribution, i.e., the denser the plasma is, the smaller will be the refraction deviation angle (Ma et al. 2010a). It may be observed from Fig. 10 that for $\theta_o = 0°$, the refraction deviation angle is also zero.

When $\theta_o = 90°$, the refraction deviation angle is straight angle, i.e., the incident EM wave propagates straightly along the tangent of the plasma cylinder. The angle at which the incident EM wave will strike cylindrical conductor will depend on plasma density, radius of conductor, and radius of plasma envelope. Since density of plasma envelope is function of distance from the center of conductor, it can be modeled as uniform multilayered unmagnetized plasma (Ma et al. 2010b). The reflection coefficient of the plasma envelope can be expressed as

$$\Gamma_o = 0 \tag{40a}$$

$$\Gamma_i = \frac{Z_{p,i+1} \cos \theta_{t,i+1} - Z_{p,i} \cos \theta_{t,i}}{Z_{p,i} \cos \theta_{t,i} + Z_{p,i+1} \cos \theta_{t,i+1}}; \; i = 1, 2, 3, \ldots, n \tag{40b}$$

$$\Gamma_n = -1 \tag{40c}$$

where

$$Z_{p,i} = \sqrt{\frac{\mu_o \mu_d}{\varepsilon_{p,i} \varepsilon_o}} = Z_o \sqrt{\frac{\mu_d}{\varepsilon_{p,i}}} \tag{41}$$

where Z_o is the free space impedance, $\varepsilon_{p,i}$ is the permittivity of ith layer of plasma, $\theta_{t,i}$ as the refraction angle at the ith layer, Γ_o is the reflection coefficient at the air to plasma interface, Γ_I is the reflection coefficient at ith to $(i + 1)$th plasma layer, and Γ_n is the reflection coefficient at nth layer to inner conductor.

Using Snell's law, (40b) can be written as

$$\begin{aligned}
\Gamma_i &= \frac{\sqrt{\varepsilon_{p,i}} \cos \theta_{t,i+1} - \sqrt{\varepsilon_{p,i+1}} \cos \theta_{t,i}}{\sqrt{\varepsilon_{p,i+1}} \cos \theta_{t,i} + \sqrt{\varepsilon_{p,i}} \cos \theta_{t,i+1}} \\
&= \frac{\sqrt{\varepsilon_{p,i}} \sqrt{1 - \frac{\sin^2 \theta_o}{\varepsilon_{p,i+1}}} - \sqrt{\varepsilon_{p,i+1}} \sqrt{1 - \frac{\sin^2 \theta_o}{\varepsilon_{p,i}}}}{\sqrt{\varepsilon_{p,i+1}} \sqrt{1 - \frac{\sin^2 \theta_o}{\varepsilon_{p,i}}} + \sqrt{\varepsilon_{p,i}} \sqrt{1 - \frac{\sin^2 \theta_o}{\varepsilon_{p,i+1}}}}
\end{aligned} \tag{42}$$

when plasma collision is considered, EM propagation through the plasma will have attenuation, i.e., reflection loss, expressed as (Ma et al. 2010b)

$$|\Gamma_{tol}|^2 = \sum_{j=1}^{n} \left\{ |\Gamma_j|^2 \prod_{q=1}^{j} \left[\left(1 - |\Gamma_{q-1}|^2\right) \exp\left(\frac{-4a_q l}{\sqrt{1 - \frac{\sin^2 \theta_o}{\varepsilon_{p,q}}}}\right) \right] \right\} \tag{43}$$

Table 15 EM propagation in cylindrical plasma envelope

Parameter		Characteristics
Incident angle (θ_o)	$\theta_o < \theta_{min}$	EM wave may be incident on the inner conductor and θ_d is $2\theta_o$
		Due to increased reflection and double-path attenuation, this case can be exploited toward refection and absorption-based stealth
	$\theta_{min} < \theta_o < 90°$	EM wave will not be incident on the inner conductor
		Higher θ_o will have lesser deviation angle
		Case can be used for refraction-based stealth
Plasma density		Higher plasma density lowers the deviation angle
Collision frequency		Increase in collision frequency decreases the reflectivity
Incident frequency		Increase in incident frequency increases the reflectivity
Index, m		Reflectivity decreases with m

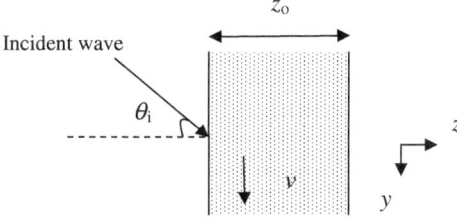

Fig. 11 Schematic for the problem of unmagnetized moving plasma

$$R_L(\text{dB}) = 10\log\left(|\Gamma_{tol}|^2\right) \qquad (44)$$

where a_q is the plasma attenuation constant of qth layer, and $l = n$.

The EM propagation characteristics in cylindrical plasma envelope (Ma et al. 2010a, b) are summarized in Table 15.

Moving plasma slab: If unmagnetized plasma is moving, the propagation characteristics of EM wave will have relativistic effect (Stanic and Okretic 1975).

If an inhomogeneous unmagnetized plasma with parabolic density profile and thickness (z_o) is assumed to be moving with uniform velocity (v) parallel to the interface (y axis), the reflection of an incident EM wave incident at an angle (θ_i) will be affected.

Figure 11 shows the schematic of the problem. The reflection behavior of moving unmagnetized plasma is summarized in Table 16.

Table 16 Reflection behavior of moving unmagnetized plasma

Angle of incidence	Performance
$\theta_i = 0$	Reflection is independent of the direction of motion
	Reflection for $|\beta| \geq 0.6$ is more than $|\beta| < 0.6$
	Reflection is independent of the plasma inhomogeneity
$\theta_i \neq 0$	Reflection is more for negative β values
	Reflection is function of β and angle of incidence
	Reflection increases with angle of incidence for $-0.6 \leq |\beta| \leq 0.6$
	TM case: Brewster's angle (θ_B) appears $-0.6 \leq |\beta| \leq 0.6$
	θ_B increases as β increases
	Plasma inhomogeneity will affect θ_B
	Reflection is function of plasma thickness ($k_o z_o$)
	For negative β, reflection increases as thickness increases and approaches to the value of one
	Plasma inhomogeneity
	Homogeneous moving plasma has slightly less (~ 10 %) reflection as compared to inhomogeneous (parabolic density profile) moving plasma
	Homogeneous plasma shows ripple-like behavior ($\beta > -0.3$) with the increase in plasma thickness, where as inhomogeneous plasma has a single peak

3 Plasma-based RCS Reduction

Plasma stealth technology is important in military applications, because of its peculiar propagation characteristics. This includes wider absorbed frequency band, attenuation, and reflection characteristics. The absorbed frequency band of plasma can be controlled by its electron density profile without altering the shape of the platform. Plasma with different parameters has different behaviors toward incident EM waves, and hence needs further research toward the realization of plasma-based stealth.

3.1 Mechanism for Plasma Generation

In general, the generation of plasma involves ionization source. The ionization source can be either electron beam, ion beam, X-rays, or UV radiation. Thus plasma density is high near the source and decreases as the distance from the source increases. Once the ionization source is switched off, plasma density decreases with time due to recombinations. At laboratory level, electron beam impact ionization and photoionization are the commonly used methods for plasma generation.

High-energy electron beam can generate plasma in air or noble gases. When the electrons get produced from the source, they diverge and interact with the back ground matter (air or noble gas). The electrons have complex spatial distribution due to backscattering, secondary electron generation, and multiple shallow angle scatterings. Electron density decreases with the distance from the source ($\sim 1/r^2$). Moreover, due to the interaction of electrons with the matter, electron energy decreases with distance. This energy deposition causes small increase in electron density near the maximum range (Vidmar 1990).

The UV sustained plasma generation technology is one of the commonly used methods for plasma generation. The photoionization of an organic vapor such as tetrakis (dimethylamino) ethylene (TMAE) seeded in background media (air or noble gas) produces collisional plasma. TMAE has low ionization potential (5.36 eV); this makes it as a preferred choice for ionization.

Gregoire et al. (1992) proposed other seed molecules for UV sustained plasma generation having independent control over the electron density and the collision process. The method uses low ionization potential seed gas (Ar) at relatively low concentration in high pressure gas (He) for UV photoionization. Here, the parameters can be optimized toward desired plasma absorption characteristics. Moreover, this method, unlike any volume ionization techniques, can decouple the plasma production and the collision process. In volume ionization techniques, the plasma is formed by the dissociation of the background gas to plasma electrons and ions.

The plasma density is proportional to derivative of the UV radiation intensity, i.e.,

$$n_e \propto -\frac{dI(x)}{dx} = \frac{I_0}{\xi} e^{-\frac{x}{\xi}}; \ I(x) = I_0 e^{-\frac{x}{\xi}} \tag{45}$$

where ς is the absorption distance, given by

$$\xi = \frac{760}{\mu P_{Ar}} \tag{46}$$

where μ is the absorption coefficient in cm^{-1} and P_{Ar} is the argon seed gas pressure in torr.

The collision frequency can be varied with the help of background gas (e.g., He) pressure. Gas purity is an important parameter to be considered. The impurities in the gas will decrease the plasma life time and the efficiency of the system.

Further, power required to maintain the electron density n_e over the volume V is given by

$$P = \frac{n_e E_i V}{\tau} = k_r n_e^2 E_i V \tag{47}$$

where E_i is the ionization potential, τ is the plasma life time, and k_r is the recombination rate.

It is well known that noble gases have low recombination rate and low affinity toward the free electrons, but have high ionization potential (10 eV or higher). This is the reason for using them as background gas in plasma generation.

The plasma cloud generated should have some kind of electromagnetically transparent enclosure. This not only keeps the plasma confined around the object but also reduces the power requirement to sustain plasma. The beam ionization involves high-energy electrons (~ 100 keV). These electrons ionize the background gas, followed by dissociation, excitation, and finally generation of plasma. The resultant plasma does not depend on the initial high-energy electrons, and the average temperature is low (<1 eV).

Plasma life time is also an important parameter that should be considered for determining the power required in sustaining the plasma. The processes that control the life time of plasma are electron–ion recombination, attachment of electrons to neutral species, and diffusion. The plasma life time is essentially a function of initial density of plasma.

The optimal design of the power supply is one of the critical issues in plasma-based RCS reduction because of its bulk weight (Chaudhury and Chaturvedi 2009). Thus, in order to avoid the heavy weight of power supply for plasma generation, it is preferred to cover only those parts of the structure which contribute toward overall RCS. Power budget calculations for the practical systems need to consider non-uniform ionization profiles created by the electron beam, and power losses in the beam generation.

3.2 Approximations and Methods: RCS Computation of Plasma Covered Shapes

After the launching of first earth satellite, it was observed that the electromagnetic properties of a body change due to the ionosphere (Swarner and Peters 1963). The collision frequency in the ionosphere is very small, and the plasma shows macroscopic properties of a dielectric lossless medium. The permittivity of plasma is even lesser than that of free space.

When a satellite moves with great velocity in ionosphere, it acquires positive or negative charge and then thus attracts the opposite charges from the ionosphere. This effect significantly alters the scattering properties of satellite. When the satellite is positively charged, it attracts electrons and may be considered as a dielectric shell with permittivity less than unity. Likewise, negatively charged satellite acquires shell of positive ions and is equivalent to a dielectric shell with permittivity greater than unity relative to the surrounding ionosphere.

This dielectric shell with permittivity less than unity can be modeled as plasma shell surrounding the objects like satellite or aircraft. The electromagnetic scattering behavior for dielectric and composite platforms can thus be estimated and analyzed using high-frequency approximation methods in conjunction with numerical techniques. The numerical technique-based methods, e.g., FDTD, are used when the medium is non-dispersive (ε, μ, and σ are constants). If the medium is frequency dependent, FDTD calculations require recursive convolution.

Chaudhury and Chaturvedi (2005) employed 3D FDTD simulation method to determine the scattering from the metallic objects shielded by the plasma shroud. Apart from multiple reflections, bending of EM wave toward the low plasma density regions has been considered to estimate the overall backscatter.

When EM wave enters into weakly ionized unmagnetized collisional plasma, it undergoes absorption and scattering. The absorption signifies the loss of wave energy, due to the transfer of energy to charged particles, and subsequently to the neutral particles during elastic and in-elastic collisions. Wave scattering takes place mainly due to spatial variation in plasma density.

In order to maximize absorption and, hence, minimize reflection, plasma density and its spatial distribution must be selected appropriately. In other words, the plasma density near the target should be sufficiently high and should fall off as the distance from the target is increased. The collision rate should also be sufficiently high. The absorption is maximized when the incident frequency matches with the collision frequency. The geometry of the circular plate shrouded by the plasma is shown in Fig. 12.

The plasma is assumed to have Epstein profile in z-direction and Gaussian profile in x–y plane. The backscattered field decreases due to the absorption and bending waves. The incident wave moves toward the low-density region due to the spatial variation in electron density of the plasma shroud. Furthermore, the incident wave that reaches the PEC plate will scatter in all directions. These scattered waves will bend in radial direction due to Gaussian profile of electron density making further reduction in the backscatter.

In the absence of plasma, maximum scattered field is in the backscatter direction and decreases slowly at other angles. However, with plasma shroud, the direction of maximum scattered field changes and fall off at other angles. This is due to the refraction effects and has significant applications toward RCS control (Chaudhury and Chaturvedi 2007).

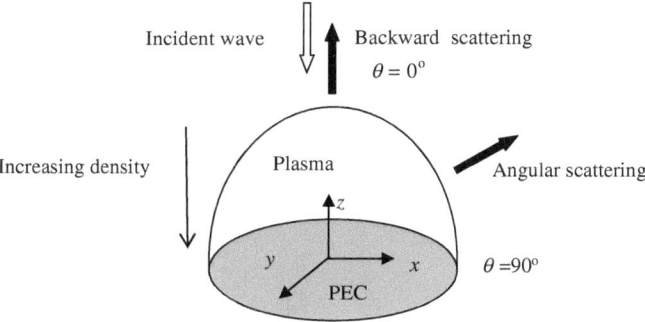

Fig. 12 A circular PEC plate with plasma shroud

Table 17 RCS of a square plate covered with unmagnetized plasma at normal incidence

Aspect angle	Bistatic RCS of a PEC square plate covered with plasma
$\theta = 0°$	Significant RCS reduction at specular direction (32.5 dB)
$0° < \theta < 90°$	(i) Significant RCS reduction (<32.5 dB) (ii) Peaks of RCS pattern shift to higher angles. This may be due to Change in electrical length through the plasma Wave bending due to plasma inhomogeneity (even for normal incidence) (iii) RCS peaks depends on Local plasma density Phase shift
$\theta = 90°$	Small increase in RCS (~ 10 dB)
$90° < \theta < 180°$	(i) Shadow region below PEC plate creates forward scattering lobe (ii) Negligible RCS reduction beyond 150°

If a metallic square plate is covered with inhomogeneous unmagnetized plasma, the effect on bistatic RCS depends not only on the aspect angle but also on the plasma covering the plate. Table 17 discusses the RCS reduction of a PEC plate for normal incidence. If the EM wave is incident at 45°, almost 20 dB RCS reduction is achieved at aspect angles $270° \leq \theta \leq 360°$, covering all specular directions (Chaudhury and Chaturvedi 2009). The variation in incident frequency, however, does not have any contribution in RCS reduction.

Another method used for analyzing plasma-based RCS reduction is *Recursive convolution finite difference time-domain* (RC-FDTD) method (Zhengli et al. 2010). If a conducting sphere is covered with unmagnetized plasma (homogeneous or inhomogeneous), greater RCS reduction is achieved in the case of inhomogeneous plasma. This is due to the fact that the homogeneous plasma has edge discontinuity, which results in strong reflections.

If conducting surface is covered with time-varying inhomogeneous, magnetized plasma, 2-D trapezoidal recursive convolution finite difference time-domain method (TRC-FDTD) can be used for analyzing bistatic RCS reduction (Liu and Zhong 2012). In the presence of an external magnetic field, plasma exhibits anisotropic behavior and the wave propagation and scattering characteristics are nonreciprocal (Geng 2011). Thus, the conventional FDTD method will not be an appropriate method for the analysis of plasma. The TRC-FDTD method can, however, be used to analyze RCS reduction of conducting surface.

The electric flux density (D) of inhomogeneous magnetized plasma can be determined in time domain using convolution integral.

In TRC method, the components of D in each FDTD cell are obtained by taking the electric field in the two consecutive time steps, given by (Liu and Zhong 2012)

$$\frac{D_x^n}{\varepsilon_o} = E_x + \sum_{m=0}^{n-1} \left[\frac{E_x^{n-m} - E_x^{n-m-1}}{2} \chi_{xx}^m - \frac{E_y^{n-m} - E_y^{n-m-1}}{2} \chi_{yy}^m \right] \qquad (48)$$

where χ_{xx}^m, χ_{xx}^m are the inverse Fourier transformations of the frequency domain permittivity functions.

Using (48), the fourth Maxwell equation can be discretized as

$$\frac{D_x^{n+1} - D_x^n}{\Delta t} = (\nabla \times H)_x^{\left(\frac{n+1}{2}\right)} \tag{49}$$

The RCS reduction due to covering of inhomogeneous, time-varying, collisional cold magnetized plasma around an infinitely long perfectly conducting cylinder (Fig. 13) with magnetic field parallel to the z-axis is summarized in Table 18.

The visual computing method (Taosheng et al. 2009) has been reported as one of the efficient methods for analyzing plasma-based RCS reduction of complex targets. In this method, laminar flow of plasma (cold, unmagnetized, collisional, and inhomogeneous) at high frequency is modeled as layered plane dielectric media.

Thus, the problem of target covered with plasma can be viewed as target covered with multilayered dielectric. The backscattering of the model (Fig. 13) can be determined using physics optics (PO) and impedance boundary condition (IBC). The flow field, along the axial direction of the target, is divided into different sections called *stations* (Fig. 14).

These stations are perpendicular to the local surface of the target. Each station flow field is then divided into layers. A number of stations and layers are determined by the plasma distribution. The backscattering is represented by scattering matrices, calculated using Fresnel reflection coefficients for the layered media.

Fig. 13 Conducting cylinder covered with inhomogeneous magnetized plasma

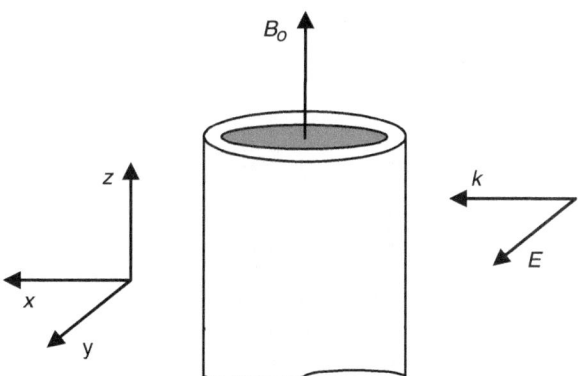

Table 18 RCS reduction in conducting cylinder covered with magnetized inhomogeneous plasma

Electron density profile	Plasma parameter	Performance
Parabolic density profile	Plasma thickness	RCS decreases as thickness increases in almost all scattering angles
		Except for ($30° < \theta < 90°$) and ($285° < \theta < 330°$)
	Cyclotron frequency	RCS increases with cyclotron frequency
		Except for ($315° < \theta < 340°$)
Time-varying Parabolic density profile	Relaxation time	Larger RCS for longer relaxation time

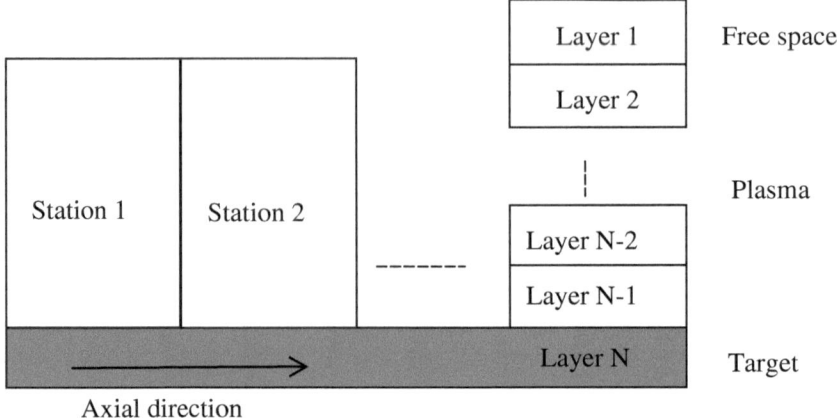

Fig. 14 Flow model of plasma covering a target

Gu et al. (2009) proposed ray tracing method for analyzing EM propagation within collisionless plasma covering the target. Non-collisional plasma has high band-pass characteristics. The EM wave attenuation is calculated by considering the refraction effect of the plasma using ray tracing method. The EM wave propagates in a straight line path passing through plasma layers. The change in the wave path in each layer will follow Snell's law.

The backscattered field of the plasma shielded target (collisionless case) is expressed as

$$E_\text{s} = 2jkpE_\text{o}\frac{e^{-jkR}}{4\pi R}\int\limits_{s}(\hat{n}_i \cdot \hat{m}_i)e^{j2k(\hat{m}_i \cdot \vec{r}_i)}\,\mathrm{d}s_i \tag{50}$$

where R is the distance between the radar and the target, k is the wave number, and i denotes the facets of the target.

It may be seen from Fig. 15 that $\vec{r}_i$ is the vector from the radar to the surface of target, $\hat{n}$ is the unit normal vector at the target surface, $\hat{m}$ is the unit vector corresponding to the wave in the innermost layer of the facet, and p is the spreading factor, given by

$$p = \sqrt{\frac{F(0)}{F(s)}} \tag{51}$$

where $F(0)$ and $F(s)$ are the wavefront sections of the ray tube at the point of incidence and on the target surface, respectively.

For collisional plasma covering a target surface, EM wave attenuation can be determined using Wentzel–Kramer–Brillouin (WKB) method (Gu et al. 2009). The reflection from the plasma boundary is neglected by assuming zero electron density

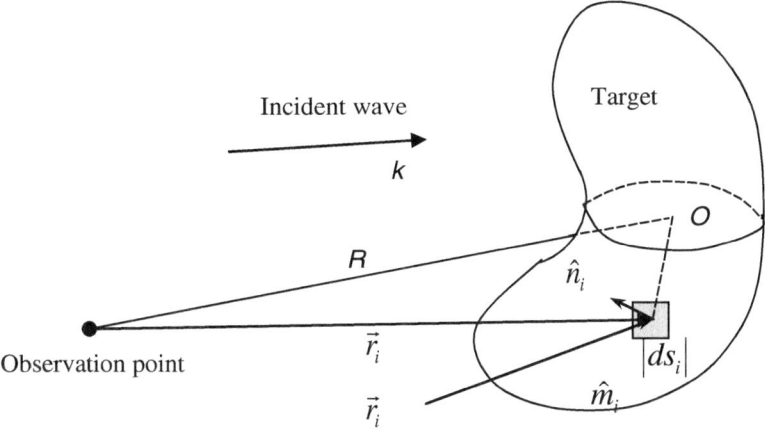

Fig. 15 Geometry for backscattered field using ray tracing

at the boundary. Plasma density is taken as layered media, which is perpendicular to gradient of plasma density, as shown in Fig. 16.

Figure 17 shows a part of metallic surface shielded with non-uniform collisional plasma. The plasma parameters are assumed to be uniform within each layer. For non-uniform plasma, according to WKB solution, a TE-polarized EM wave is expressed as

$$E_y = E_o \exp \left[\mp \int_0^z \sqrt{k^2(z) - k_o^2 \sin^2 \theta_n} \, dz \right] \tag{52}$$

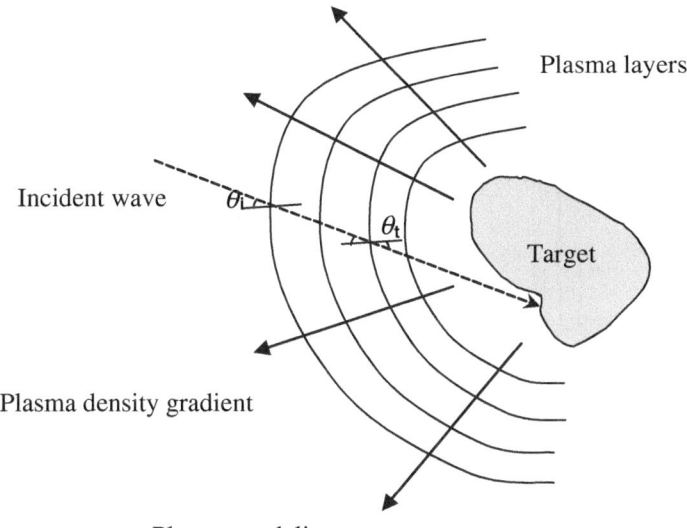

Fig. 16 Plasma modeling

Fig. 17 EM wave
propagating in through
non-uniform plasma

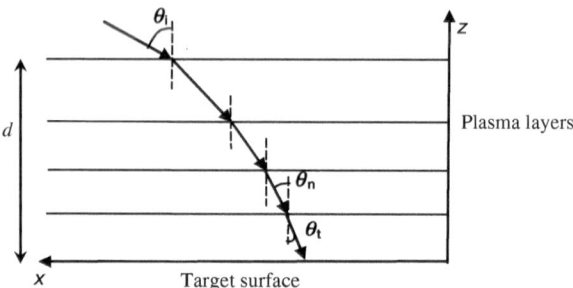

EM wave undergoes a two-way attenuation while traveling through the plasma layers due to reflection and transmission. This attenuation (in dB) is expressed as

$$\text{Att} = \left| 10 \log \frac{P_R(z_0)}{P_0} \right| = \left| 17.37 \, \text{Im} \int_0^{z_0} \sqrt{k^2(z) - k_0^2 \sin^2 \theta_n} dz \right| \qquad (53)$$

where P_0 is the incident power and $P_R(z_0)$ is the reflected power at z_0, given by

$$P_R(z_0) = P_0 \exp \left(-4 \, \text{Im} \int_0^{z_0} \sqrt{k^2(z) - k_0^2 \sin^2 \theta_n} dz \right) \qquad (54)$$

If a sphere is coated with inhomogeneous unmagnetized plasma (Fig. 18), the RCS reduction depends directly on the plasma density (Taosheng et al. 2009), when collision frequency is same as incident wave frequency. This is true even when electron density and collision frequency have parabolic distribution in radial direction (Gu et al. 2009), i.e.,

$$n_e(r) = n_{eo} \left[1 - \frac{(r - r_0)}{d^2} \right]$$

$$v(r) = v_0 \left[1 - \frac{(r - r_0)}{d^2} \right] \qquad (55)$$

Fig. 18 Sphere shielded with
plasma

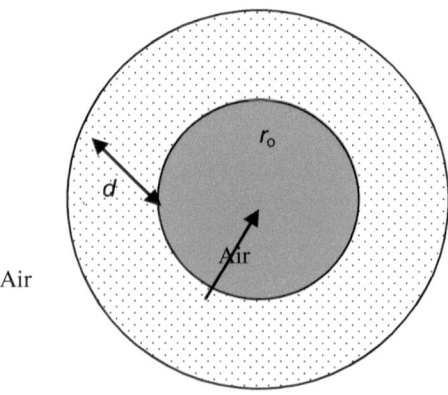

where r is the radius of each layer $\{r_o \leq r \leq (r_o + d)\}$, and n_{eo} and v_o are the electron density and the collision frequency in the innermost layer, respectively. This trend of plasma-based RCS reduction may also be observed for conical targets and dihedral corner reflectors (Ruifeng and Donglin 2003). In dihedral corner reflectors, the RCS reduction up to 30-40 dB can be achieved.

The conductivity of the surface is one of the important parameters that can be exploited toward the control of scattering. If an incident EM wave propagates in a medium having zero conductivity (e.g., air), the skin depth would determine the volume over which scattering is effective. In other words, volume scattering can be approximated as surface scattering (Blackledge 2007).

If plasma covers the scatterer surface, it will avoid sharp variation in conductivity at the air–scatterer interface. In other words, the plasma will cause absorption of the incident EM wave before it is incident on the conducting scatterer. The plasma essentially creates a conductive shield that protects the object from incident EM wave. At a given frequency, skin depth depends on the plasma conductivity. The more the plasma is conducting, the smaller will be the skin depth. At higher frequencies, the penetration of incident EM wave will be very less. This is true for highly conductive plasma. However, in case of weakly conductive plasma, the effect of plasma will be dominated by the exponential factor, $\exp\left(-\frac{\sigma_o t}{\varepsilon_o}\right)$, where σ_o and ε_o are the conductivity and permittivity of plasma. This exponential factor has great significance. For larger conductivity, the plasma effect on reduced scattering degrades exponentially. Moreover, the scattered field is independent of frequency. It may be noted that in practical applications, plasma used for screening is weakly ionized and weakly conductive.

For weakly ionized plasma, the conductivity depends on the electron density. Thus, the main factors that affect the plasma performance are plasma density, the plasma thickness, and the continuity at the air–plasma interface. These three factors in turn depend on the power of the plasma, stability of the plasma, and its profile. Thus, the electron density profile of the plasma is the parameter, which needs to be optimized toward plasma shielding effect, and hence RCS reduction of aerospace structures.

The conductivity of plasma depends on the extent of ionization. For weakly ionized plasma, the conductivity is given by

$$\sigma \approx \frac{ne^2}{m_e v_{ea}} \tag{56}$$

where v_{ea} is the collision frequency between electron and atom, and n is the electron number density. The ratio n/v_{ea} varies as aerospace vehicle moves from one region (altitude and speed of flight) to other region. If hydrogen atoms are produced prior to ionization, then it is possible to maintain large electron densities with low collision frequencies. This in turn leads to large and stable plasma conductivities. This enhances the effectiveness of plasma screening. Thus, an important issue is to determine the electron density distribution for a given configuration of the plasma source and aerospace platform. Moreover, parameters like velocity of vehicle, plasma medium, additives like water vapor, electron beam energy, its diameter, and profile will determine the extent of plasma screen.

3.3   Plasma with RAM Material

The plasma shielding effect can further be improved by coating the aerospace vehicle with *radar absorbing material* (RAM). The purpose is to absorb the portion of incident EM wave that manages to propagate through the plasma layer surrounding the vehicle. The propagation characteristics of EM wave in a multilayered *radar absorbing structure* (RAS) consisting of plasma and RAM have been studied (Table 19).

Table 19   Performance of the target covered with plasma and RAM

Parameters	Performance of the structure
Incident wave frequency	*Low frequency*: Target with RAM shows better RCS reduction than the target covered with plasma and RAM
	Reason: EM waves are partially absorbed and reflected within plasma layer, before reaching RAM
	High frequency: Target with plasma and RAM shows better RCS reduction
	Reason: EM waves are absorbed by both plasma and RAM
Interface $z = d_1$	Interface at $z = d_1$ results in reflections, thereby decreasing the stealth performance
	Reflections at $z = d_1$, increase the overall reflection from the structure. This reduces the absorption within the plasma and RAM
Plasma parameters (Layer 2)	*Plasma thickness*: Bounded plasma shows nonlinear relation between attenuation and plasma thickness
	Attenuation occurs due to cavity resonance effect of the bounded plasma and the collisions. However, the attenuation peaks are due to cavity resonance effect
	If the plasma thickness is too high, attenuation becomes independent of the plasma thickness Plasma thickness needs to be optimized to obtain stronger cavity resonance effect
	Plasma electron density: Attenuation peaks are shifted to higher frequencies as electron density increases
	Plasma collision frequency: It will affect the attenuation peaks and its position
	Increase in collision frequency will increase the EM wave absorption
RAM parameters (Layer 3)	At higher frequencies, RAM parameters should be matched to plasma parameters to get more EM wave absorption
Lossless transparent material (Layer 1)	Dielectric constant of Layer 1 should be selected appropriately so as to avoid the mismatch between the plasma and air

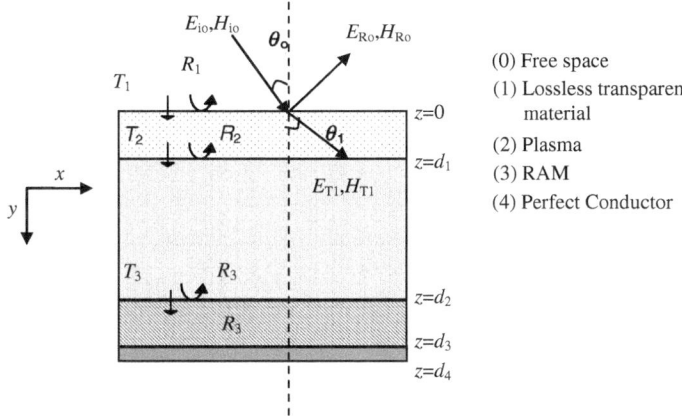

Fig. 19 Wave propagation in a multilayered structure consisting of plasma and RAM

Chaohui et al. (2008) employed FDTD to analyze the interaction of EM waves with the plasma and RAM covered conducting plate. It is reported that plasma covering the RAM-coated metallic plate shows relatively high-frequency absorption as compared to plasma covering plate or RAM-coated plate. If RAM is placed over the plasma-covered plate, the absorption takes place at lower frequencies. The optimization of plasma and RAM parameters is expected to enhance the absorption bandwidth.

Yuan et al. (2011) proposed a four-layered structure consisting of a PEC, absorbing material, plasma, and lossless transparent material (Fig. 19) toward RCS reduction using both plasma and RAM. An impedance transformation approach has been used to analyze the performance of the structure in terms of reflectance and attenuation.

4 Challenges in Plasma-based RCS Reduction

The realization of plasma-based stealth is undoubtedly a challenge for engineers and researchers. One main concern is that the plasma cloud surrounding the structure may give rise to other signatures such as thermal, acoustic, infrared, or visual from the structure (Vass 2003). The technical difficulties in realizing the plasma-based RCS reduction should be considered while designing a stealthy structure. Such issues (Cadirci 2009; Yuan et al. 2011) are outlined below:

(i) Plasma might give rise to EM radiation with visible glow due to neutralization of plasma ions. One way to avoid such optical signature is to add an opaque plastic enclosure. However, this might not be feasible for many situations especially for aircraft. This type of enclosure should be movable and adjustable according to the requirements (Chaudhury and Chaturvedi 2009). They should be transparent to EM waves and should not induce scattering. Moreover, enclosure should not also disturb the onboard avionics. Ma et al. (2008) proposed a glass-reinforced plastic envelop with quarter-wavelength height for reducing the reflections from the enclosure.

(ii) Plasma trail of ionized layer behind the aircraft before dissipation into the atmosphere
This type of trail may result in both visible and radar signatures. This also necessitates an enclosure to avoid the plasma trail. Alternatively, the dissipation rate of plasma trail should be increased. If aircraft fly at high altitudes or operates during day time, plasma trail problem can be avoided (http://www.stealthskater.com).

(iii) Generation of radar absorbent plasma around the aircraft moving with high speed

(iv) Requirement of windows in plasma fields for aircraft onboard avionics.
This issue can be treated by switching off the plasma generator at the time of communication.

(v) Power required to sustain the plasma around the aircraft
The plasma around the aircraft should be sustained so as to compensate the atomic processes such as recombination and electron attachment. This challenge necessitates the proper choice of the plasma density profiles toward the optimum input power required for plasma generation and maintenance.

(vi) Mechanisms for temperature stability
Temperature should be maintained with the help of proper cooling mechanisms. The high temperature of plasma cloud surrounding the aircraft might harm the aircraft itself. Furthermore, it might give rise to infrared signatures. Mechanism of plasma generation should be designed in such a way that it should not increase the gas temperature.

(vii) Weight of the plasma generator
Plasma is light weight but its generator is bulky and heavy. Moreover, the power required for plasma generation is high. The solution for this issue is to keep plasma generator switched on only when required (Singh et al. 2005). Furthermore, instead of covering whole aircraft structure, only certain parts that contribute significantly toward RCS should be covered with plasma layer.

(viii) Lack of availability of experimental data limits the research in analyzing the design and realization of plasma-based stealthy structures.

5 Conclusion

Plasma has been considered as one of the means for controlling the RCS of aircraft-like structures. Although limited details and data related to plasma-based RCS reduction and control are available, it is known that a few countries have implemented plasma stealth in their warfare.

The EM wave interaction with plasma mainly depends on the physical characteristics and associated plasma parameters, especially the plasma temperature and the plasma density. Plasma is electrically conductive over wide range of frequencies. Plasma acts as reflector to low frequencies. The use of plasma to control the reflected EM wave and, hence, the RCS is possible at higher frequencies, where the conductivity of the plasma results in greater interactions of plasma and the incident EM wave. The wave is absorbed within the plasma and thus contributes toward RCS reduction.

For stealth application, both unmagnetized and magnetized weakly ionized collisional plasmas have been analyzed. In magnetized plasma, the absorption of incident EM wave is essentially due to the cyclotron resonance, whereas in unmagnetized, plasma absorption of EM wave is due to the collision of electrons. The plasma frequency and, hence, electron density plays significant role in both magnetized and unmagnetized plasma. It is inferred that the plasma performance toward the RCS reduction and control depends on its parameters and conditions in which it is used as a covering to a structure.

Table 20 summarizes the role of plasma parameters in EM wave propagation characteristics. The propagation characteristics of plasma can be studied using scattering matrix method or time recursive convolution method in conjunction with FDTD. The transformation of EM wave polarization when it propagates through plasma can be dealt efficiently by numerical technique-based approaches such as FDTD or propagator matrix method (PMM). This is due to requirement of high computational time and memory. Table 21 presents maximum RCS reduction in case of different shapes when plasma is used to cover them.

Furthermore, the plasma generators are bulkier. Thus, instead of covering whole aircraft structure with plasma cloud, only those parts which are significant contributors to overall RCS can be covered with plasma. These include antennas, dihedral like structures, edges, etc. This will optimize the power supply of plasma generator as well the bulkiness. Moreover, the power supply of plasma generator need not be switched on continuously. Emissions from the probing radars can also be utilized to activate the plasma generation.

The future of the plasma lies not only in plasma-based screening but also in the area of plasma-based antennas and frequency selective surface (FSS)-based structures. Since antennas are significant contributor to the platform RCS, they can be designed using plasma instead of metals. Similarly, the metallic portions of FSS can be replaced by plasma, thereby reducing the scattering cross section.

Table 20 Role of plasma parameters

Parameter	Performance
Electron density (n_p)	Increase in n_p and, hence, ω_p
	• Reflected and absorbed power increase (*for magnetized plasma*)
Plasma frequency (ω_p)	Frequency at which maximum occurs increases for both parabolic and exponential density profiles
	Bandwidth increases for parabolic case while remains same for exponential case
	Absorption band increases for linear and linear plus sinusoidal electron density profile
	• Transmitted power decreases; Increases beyond the cyclotron frequency (ω_c) (*for magnetized plasma*)
	Frequency at which maximum occurs increases
	Bandwidth increases
	• Lower deviation angle for the incident wave (for *unmagnetized* plasma)
Collision frequency (v)	Increase in v
	• Reflected and absorbed power decrease (*for magnetized plasma*)
	Frequency at which maximum occurs remains same for parabolic, linear and exponential density profile
	Bandwidth remains same for parabolic and exponential density profile while increases for linear and linear plus sinusoidal electron density profile
	• Transmitted power increases; Decreases after the cyclotron frequency (*for magnetized plasma*)
	Frequency at which maximum occurs remains same
	Bandwidth slightly increases for parabolic and exponential density profile
Other parameters	• Absorption is due to the collisions and cavity resonance effect (*for unmagnetized plasma*)
Plasma thickness	
Incident frequency	At high frequencies, cavity resonance effect dominates over the collisional absorption and the absorption and reflection increases with increase in plasma thickness
Plasma temperature	
	If the plasma thickness is too high, attenuation becomes independent of plasma thickness
	Optimum plasma thickness for efficient reflection characteristics
	• Increase in incident frequency results in reduction of reflected power (*for unmagnetized plasma*)
	Total reflection angle and the reflection loss increases
	• Increase in temperature (*for unmagnetized plasma*)
	Collision frequency increases
	Reflection loss and absorption increases
Electron density profile	• Parabolic electron density profile has greater absorbing and reflecting powers than the exponential (*for magnetized plasma*)
	• Exponential density has greater transmitting power and wide absorption band than parabolic profile (*for unmagnetized plasma*)
	• Partially linear and sinusoidal profile will provide better matching at the plasma–boundary
	Improved absorption and transmission characteristics (*for unmagnetized plasma*)
	Depends on applied magnetic field (*for magnetized plasma*)

(continued)

Table 20 (continued)

Parameter	Performance
Cyclotron frequency (ω_c)	• It can tune the absorption band • Absorption band will move from low-frequency to high-frequency region as the magnetic field strength increases, without changing the bandwidth • Less reflection, more absorption, and zero transmission near the cyclotron frequency • When $\omega_c \gg \omega_p$, peak reflection and absorption are very close to cyclotron frequency
Incidence angle	*For magnetized plasma* • Absorption band widens as the incident angle w.r.t. the plasma slab increases *For unmagnetized plasma* • Reflection loss is maximum for one particular incidence angle (for TE wave) • Reflection loss decreases as incident angle increases (for TM wave) • At normal incidence, TE and TM have same reflection loss • For oblique incidence, TM has less reflection loss than TE wave

Propagation characteristics *(for unmagnetized plasma)*
• Absorption by the plasma is due to the collisions, which leads to RCS reduction
• Wave scattering is due to spatial variation in plasma density
• Refraction of incident EM wave determines the extent of backscatter

RCS reduction increases by
• Increasing the collision frequency *(for both magnetized and unmagnetized plasma)*
• Increasing the electron density *(for unmagnetized plasma)*
• Longer the relaxation time of ionized plasma *(for magnetized plasma)*

Table 21 RCS reduction reported for different shapes

Shape of PEC target	Plasma medium	RCS reduction (maximum) (dB)
Square plate	Inhomogeneous unmagnetized	>20
		32.5 (for normal incidence)
Dihedral corner reflectors	Unmagnetized	30–40
Sphere	Homogeneous unmagnetized	10
	inhomogeneous unmagnetized	30
Cylinder	Inhomogeneous magnetized	22
Aircraft (Su-27IB)		20

References

Alexef, I., W.L. Kang, M. Rader, C. Douglass, D. Kintner, R. Ogot, and E. Norris. 1998. A plasma stealth antenna for the U.S navy. In *Proceedings of IEEE International Conference on Plasma Science,* Raleigh, NC, USA, 1 pp. 1–4 June 1998.

Anderson, T., S. Parameswaran, E.P. Pradeep, J. Hulloli, and P. Hulloli. 2006. Experimental and theoretical results with plasma antennas. *IEEE Transactions on Plasma Science* 34(2): 166–172. April 2006.

Beskar, C.R. 2004. Cold plasma cavity active stealth technology," Technical White paper, Stavatti Military Aerospace: Tactical Air Warfare Systems Division, South St. Paul MN, USA, 11 p., November 2004.

Blackledge, J.M. 2007. Modeling and computer simulation of radar screening using plasma clouds. *ISAT Transactions on Electronics and Signal processing* 1(1): 61–71. January 2007.

Born, M., and E. Wolf. 2002. In *Principles of Optics; Electromagnetic Theory of Propagation, Interference and Diffraction of Light.* 7th edn. Cambridge, UK: Cambridge University Press, ISBN: 0521642221, 952 pp.

Cadirci, S. 2009. RF Stealth (low observable) and counter RF stealth technologies: Implications of counter RF stealth solutions for Turkish Air Force. Naval Postgraduate School, Monterey, California, Master's Thesis Report, 161 pp., March 2009.

Chaohui, L., H. Xiwei, and J. Zhonghe. 2008. Interaction of electromagnetic waves with two-dimensional metal covered with radar absorbing material and plasma. *Plasma Science and Technology* 10(6): 717. 2008.

Chaudhury, B., and S. Chaturvedi. 2007. Radar cross section reduction using plasma blobs: 3D finite difference time domain simulations. In *Proceedings of IEEE Applied Electromagnetic Conference*, Kolkata, 4 pp., 19–20 December 2007.

Chaudhury, B., and S. Chaturvedi. 2009. Study and optimization of plasma based radar cross section reduction using three dimensional computations. *IEEE Transactions on Plasma Science* 37(11): 2116–2127. January 2007.

Chaudhury, B., and S. Chaturvedi. 2005. Three dimensional computation of reduction in radar cross section using plasma shielding. *IEEE Transactions on Plasma Science* 33(6): 2027–2034. December 2005.

Chawla, B.R., and H. Unz. 1969. Reflection and transmission of electromagnetic waves normally incident on a plasma slab moving uniformly along a magnetostatic field. *IEEE Transactions on Antennas and Propagation* 17(6): 771–777. November 1969.

Chen, F.F. 1974. *Introduction to plasma physics.* New York: Plenum press, ISBN: 0-306-30755-3, 329 pp.

Dinklage, A. 2005. *Plasma physics: confinement, transport and collective effect.* New York: Springer, ISBN: 3540252746, 496 pp.

Froula, D.H., S.H. Glenzer, N.C. Luhmann, Jr., and J. Sheffield. 2011. *Plasma Scattering of Electromagnetic Radiation: Theory and Measurement Techniques.* 2nd edn. Burlington, MA, USA: Elsevier, Academic Press, ISBN: 9780123748775, 497 pp.

Geng, Y.L. 2011. Scattering of plane wave by an anisotropic plasma-coated conducting sphere. *International Journal of Antennas and Propagation* 2011:6. Article Id 409764. 2011.

Ginzburg, V.L. 1961. *Propagation of electromagnetic waves in plasma.* New York: Gordon and Breach Science Publishers, ISBN-10: 0677200803, 822 pp.

Gregoire, D.J., J. Santoru, and R.W. Schumacher. 1992. Electromagnetic wave propagation in unmagnetized plasmas. Technical Report, Hughes Research Labs, Malibu, CA, 66 pp., March 1992.

Gruel, C.S., and E. Oncu. 2009. Interaction of electromagnetic wave and plasma slab with partially linear and sinusoidal electron density profile. *Progress in Electromagnetic Research Letters* 12: 171–181. 2009.

Gu, W., Y. Lei, W. Taosheng, F. Ning, M. Jungang, and W. Baofa. 2009. "RCS calculation of complex targets shielded with plasma based on visual GRECO method. In *Proceedings of*

International Symposium on Microwave Antenna Propagation and EMC Technologies for Wireless Communications, Beijing, pp. 950–953, Oct 2009.

Hu, B.J., G. Wei, and S.L. Lai. 1999. SMM analysis of reflection, absorption and transmission from nonuniform magnetized plasma slab. *IEEE Transactions on Plasma Science* 27(4): 1131–1136. August 1999.

Jenn, D.C. 2005. *Radar and Laser Cross Section Engineering*. 2nd ed., AIAA Education Series, Washington DC, ISBN-13: 9781563477027, 505 pp.

Laroussi, M., and J.R. Roth. 1993. Numerical calculation of the reflection, absorption and transmission of microwaves by nonuniform plasma slab. *IEEE Transactions on Plasma Science* 21(24): 366–372. August 1993.

Liu, S., and S. Zhong. 2012. FDTD study on scattering for conducting target coated with magnetized plasma of time varying parabolic density distribution. *Progress in Electromagnetics Research M* 22: 13–25. January 2012.

Ma, L.X., H. Sang, L. Zhu, and X.J. Gao. 2010a. Analysis on the refraction stealth characteristics of cylindrical plasma envelopes. In *Proceedings of International Conference on Microwave and Millimeter wave Technology*, Chengdu, pp. 1695–1698, May 2010.

Ma, L.X., H. Zang, Z. Li, and C.X. Zhang. 2010b. Analysis on the stealth characteristics of two dimensional cylinder plasma envelopes. *Progress in Electromagnetic Research Letters* 13: 83–92. 2010.

Ma, L.X., H. Zhang, and C.X. Zhang. 2008. Analysis on the reflection characteristic of the electromagnetic wave incidence in closed non magnetized plasma. *Journal of Electromagnetic waves and Applications* 22(17–18): 2285–2296. 2008.

Mo, J., and N. Yuan. 2008. Analytical solution of reflection coefficient microwaves oblique incidence on a nonuniform magnetized plasma slab. *International Conference on Microwave and Millimeter wave Technology, Nanjing* 4: 1930–1933. April 2008.

Roth, J.R. 1994. Interaction of electromagnetic fields with magnetized plasmas. Scientific Report PSL-94-3, UTK Plasma Science Laboratory, University of Tennessee, Knoxville, TN, 329 pp., March 1994.

Ruifeng, L., and S. Donglin. 2003. Emulation research about feasibility of reducing dihedral corner reflector RCS with plasma. In *Asia Pacific Conference on Environmental Electromagnetics,* Hangzhou, China, pp. 523–526, Nov 4–7, 2003.

Sadeghikia, F., and F.H. Kashani. 2013. A two element plasma antenna array. *ETASR-Engineering, Technology & Applied Science Research* 3(5): 516–521. 2013.

Seshadri, S.R. 1973. *Fundamentals of Plasma Physics.* American Elsevier Publisher, New York, ISBN: 0-444-00125-5, 545 pp.

Singh, A.K., B.S. Bhadoria, A.K. Kushwaha, and K. Chaturvedi. 2005. *Scope and Challenge in Plasma: Science & Technology*. Allied Pub, New Delhi, ISBN: 81-77648659, 141 pp.

Singh, Y.P., and A.S. Shekhawat. 1983. Interaction of obliquely incident electromagnetic wave with collisional, magnetized and moving plasma slab. *Acta Physica Hungarica* 54: 101–109. 1983.

Skolnik, M.I. 2003. *Introduction to Radar Systems,* 3rd edn. New York: Tata McGraw-Hill Education, ISBN: 0070445338, 772 pp.

Stanic, B.V., and V.K. Okretic. 1975. Reflection of electromagnetic waves by a moving ionized layer with parabolic electron density profile. *Univ. Beograd. Purl. Elektrotehn. Fak.* 15: 225–234. 1975.

Swarner, W.G., and L. Peters. 1963. Radar cross sections of dielectric or plasma coated conducting spheres and circular cylinders. *IEEE Transactions on Antennas and Propagation* 11(5): 558–569. September 1963.

Taosheng, W., Y. Lei, W. Gu, F. Ning, and W. Baofa. 2009. Visual computing method of radar cross section for target coating with plasma. *Chinese Journal of Electronics* 18(3): 579–582. July 2009.

Vass, S. 2003. Stealth technology deployed on the battle field. *Informatics Robotics* 2(2): 257–269. 2003.

Vidmar, R.J. 1990. On the use of atmospheric pressure plasmas as electromagnetic reflectors and absorbers. *IEEE Transactions on Plasma Science* 18(4): 733–741. August 1990.

Williams, E.R., and S.G. Geotis. 1989. A radar study of the plasma and geometry of lightning. *Journal of the Atmospheric Sciences* 46(9): 1173–1185. May 1989.

Yin, X., H. Zhang, S. Sun, Z. Zhao, and Y. Hu. 2013. Analysis of propagation and polarization characteristics of electromagnetic waves through the nonuniform magnetized plasma slab using propagator matrix method. *Progress in Electromagnetic Research* 137: 159–186. 2013.

Yu, Z., Z. Zhang, L. Zhou, and W. Hu. 2003. Numerical research on the RCS of plasma. In *International Symposium on Antennas, Propagation and EM Theory*, Beijing, China, pp. 428–432, Oct, 28–Nov, 1, 2003.

Yuan, C.X., Z.X. Zhou, and H.G. Sun. 2010. Reflection properties of electromagnetic wave in a bounded plasma slab. *IEEE Transactions on Plasma Science* 38(12): 3348–3355. December 2010.

Yuan, C.X., Z.X. Zhou, J.W. Zhang, X.L. Xiang, Y. Feng, and H.G. Sun. 2011. Properties of propagation of electromagnetic wave in a multilayer radar absorbing structure with plasma and radar absorbing material. *IEEE Transactions on Plasma Science* 39(9): 1768–1775. September 2011.

Zhengli, H., J. Ding, P. Chen, Z. Zhang, and C. Guo, FDTD analysis of three dimensional target covered with inhomogeneous unmagnetized plasma. In *International Conference on Microwave and Millimeter wave Technology*, Chengdu, pp. 125–128, May 8–11, 2010.

About the Book

This book presents a comprehensive review of plasma-based stealth, covering the basics, methods, parametric analysis, and challenges towards the realization of the idea. The concealment of aircraft from radar sources, or stealth, is achieved through shaping, radar absorbing coatings, engineered materials, or plasma, etc. Plasma-based stealth is a radar cross section (RCS) reduction technique associated with the reflection and absorption of incident electromagnetic (EM) waves by the plasma layer surrounding the structure. A plasma cloud covering the aircraft may give rise to other signatures such as thermal, acoustic, infrared, or visual. Thus it is a matter of concern that the RCS reduction by plasma enhances its detectability due to other signatures. This needs a careful approach towards the plasma generation and its EM wave interaction. The book starts with the basics of EM wave interactions with plasma, briefly discusses the methods used to analyze the propagation characteristics of plasma, and its generation. It presents the parametric analysis of propagation behaviour of plasma, and the challenges in the implementation of plasma-based stealth technology. This review serves as a starting point for the graduate and research students, scientists and engineers working in the area of low-observables and stealth technology.

© The Author(s) 2016   47
H. Singh et al., *Plasma-based Radar Cross Section Reduction*,
SpringerBriefs in Computational Electromagnetics,
DOI 10.1007/978-981-287-760-4

Author Index

© The Author(s) 2016
H. Singh et al., *Plasma-based Radar Cross Section Reduction*,
SpringerBriefs in Computational Electromagnetics,
DOI 10.1007/978-981-287-760-4

Subject Index

© The Author(s) 2016
H. Singh et al., *Plasma-based Radar Cross Section Reduction*,
SpringerBriefs in Computational Electromagnetics,
DOI 10.1007/978-981-287-760-4